The Cognitive Revolution

The Cognitive Revolution

How AI Is Reorganizing Intelligence, Expertise, and Institutions

Jiajie Zhang

Open Intelligence Press

Published by Open Intelligence Press

First edition

ISBN: 979-8-9956147-2-2

The views expressed in this book are solely those of the author.

To those who know intelligence has already changed—and are ready to redesign what follows.

For those who refuse to inherit a world built for a form of intelligence that no longer exists.

"We shape our tools, and thereafter our tools shape us."

— Marshall McLuhan

The Cognitive Revolution

How AI Is Reorganizing Intelligence, Expertise, and Institutions

TABLE OF CONTENTS

Intelligence Has Left the Building

Intelligence has left the building.

For centuries, we believed it lived inside the human mind—trained through education, measured through exams, and validated through credentials. We built entire institutions on that belief.

That belief is now obsolete.

This book begins with a question that no longer has the answer we were taught:

Where does intelligence reside?[1]

I Was There at the Beginning—But We Didn't Know It

In the late 1980s, as a doctoral student at the University of California, San Diego, I entered a rare intellectual environment—one that, in retrospect, was extraordinary and helped shape the current moment.

The Institute for Cognitive Science brought together an interdisciplinary approach to intelligence across brain, behavior, and computation. Under the leadership of Donald A. Norman and David E. Rumelhart, it was not just studying intelligence. It was redefining it.[2]

One area of that effort was Parallel Distributed Processing, or PDP.

The claim was radical.

Intelligence did not have to be built from discrete symbols and rules.

It could emerge from patterns distributed across simple units—neurons—operating in parallel.

The breakthrough that made this operational was the backpropagation, developed by Rumelhart, Hinton, and Williams.[3]

> A learning algorithm that allowed networks to adjust themselves.
> To learn.
> To improve.
> To scale.

This is one of the foundational algorithms of modern AI. I was an early enthusiast of PDP, serving as the first teaching assistant for the undergraduate course built on this paradigm, where I taught backpropagation in LISP as these ideas were taking shape.
I did not ultimately build my career inside neural networks. But I stayed close to the field—shadowing its evolution as it moved from a promising theory to the core infrastructure of modern AI.
Instead, I moved into a different but related line of inquiry—one that asked a broader question.

> If intelligence can be distributed across neurons, what else can it be distributed across?

From Neurons to Systems

Led by Don Norman and Ed Hutchins, the Distributed Cognition Group extended the idea of distribution beyond the brain.[4]
Not just across neurons.
But across:

> people
> artifacts
> environments
> space and time

The implication was profound.
Thinking is not inside the head.
It is a system-level phenomenon.[5]

What PDP revealed computationally,
distributed cognition made conceptually explicit.

One asked how intelligence could emerge from parallel
processing in neural networks.
The other asked how cognition emerges in socio-technical
systems.

Two ideas.
Born in the same intellectual home.
Then moving along different paths.

Two Trajectories

The first trajectory was neural networks.
After its early breakthroughs, the field stalled.

The AI winter arrived.
Momentum slowed.

Neural networks became an idea ahead of its infrastructure.
Then, in 2012, everything changed.
Geoffrey Hinton and his team achieved a breakthrough in
the ImageNet competition.[6]
Deep neural networks did not merely improve performance.
They reset the baseline.
Machines began outperforming previous methods—and, on
some perceptual tasks, even humans.
Neural networks returned, no longer as a promising theory,
but as a practical and dominant technology at industrial
scale.
The second trajectory—distributed cognition—never
entered a winter.

It continued to develop quietly.

> Not as a dominant technology,
> but as a framework.

A way to understand how thinking actually happens:

> across teams
> across tools
> across workflows
> across time

It did not scale through computation.
It matured through explanation.
It became a powerful way to understand socio-technical systems long before AI made that understanding unavoidable.
For decades, these two trajectories evolved in parallel.
One advanced the computational power of intelligence.
The other expanded our understanding of where intelligence resides.
Only recently have they converged.

And Then It Happened

AI crossed a threshold.

> Not when it became faster.
> Not when it became more accurate.

But when it began to participate in thinking.[7]
With large language models, something fundamental changed.
We no longer simply use tools.
We think with systems that respond.

We write through interaction
We reason through dialogue
We explore through iteration

Language is no longer just expression.
It is computation.
Thought is no longer only internal.
It is interactive.

This is not an upgrade.
This is a relocation.

The Gap I Could Not Ignore

As this shift accelerated, I saw a widening gap.
On one side:

rapidly advancing AI capability
expanding cognitive reach
new forms of reasoning

On the other:

slow-moving institutions
unchanged structures
superficial adaptation

There was activity everywhere.
But no redesign.
I kept returning to the same conclusion:

Many organizations are not transforming.
They are performing transformation.

Because they are solving the wrong problem.
They are adding AI.
Instead of redesigning cognition.[8]

The Claim

This book makes a simple claim:

> Intelligence is no longer an individual property.
> It is a system property.[5]

AI did not create this shift.
It revealed it.
And accelerated it.
Once intelligence becomes distributed in practice,
everything built on individual cognition becomes unstable:

> expertise
> education
> research
> decision-making
> governance
> institutions

Unless they are redesigned.

What This Book Is About

This is not a book about AI.
It is a book about intelligence.
And more importantly:
It is a book about designing intelligence.
Because distributed cognition does not organize itself well.
Without design:

> systems fragment
> decisions degrade
> alignment breaks

With design:

cognition scales
learning accelerates
institutions transform

Why This Matters to Me

I have lived across both trajectories:

the rise, dormancy, and return of neural networks
the development of distributed cognition as a framework
for understanding socio-technical systems
What one built computationally,
the other explained conceptually.

Now they have converged.
What was once theory is now infrastructure.
What was once optional is now inevitable.
The risk is not that we misunderstand AI.
It is that we misunderstand what it means for intelligence to
move.

If You Take One Idea From This Book

Take this:
The brain is no longer the boundary of intelligence.
Everything else follows.

The Invitation

This book does not explain AI.
It challenges how you think about thinking.
If it works, you will see:

cognition as a system
institutions as cognitive architectures
AI as infrastructure

And you will recognize:

We are not adapting to technology.
We are redesigning cognition.

Closing

The cognitive revolution is already here.
Not in the future.
In everyday thinking.
The only question is:
Will we design for it—or be constrained by what we refuse
to redesign?
This book is an argument for design.
And an invitation to begin.

Endnotes

1. This opening claim draws on distributed and extended cognition. Edwin Hutchins's Cognition in the Wild (1995), Donald A. Norman's The Design of Everyday Things (1988), and Andy Clark and David Chalmers's "The Extended Mind" (1998) argue, in different ways, that thinking is not sealed inside the skull but unfolds through coordination with artifacts and environments. Even reading is better understood as an interaction between reader and text than as a purely internal event.
2. The bridge from neural distribution to socio-technical distribution runs across two literatures. David Rumelhart and James McClelland's Parallel Distributed Processing (1986) shows distribution across neurons within a computational system; Donald A. Norman's Things That Make Us Smart (1993) and Edwin Hutchins's Cognition in the Wild (1995) extend the same intuition outward to people, artifacts, and settings. That is why the two trajectories in this Preface belong in the same story.
3. For the specific breakthrough on backpropagation, the key source is David Rumelhart, Geoffrey Hinton, and Ronald Williams's "Learning Representations by Back-Propagating Errors" (1986). For the wider PDP framework in which it mattered, see Rumelhart and McClelland's Parallel Distributed Processing (1986). Together they showed that useful internal representations could be learned rather than hand-coded.

4. The most direct source for this socio-technical turn is Edwin Hutchins's Cognition in the Wild (1995). For how external representations themselves carry part of the cognitive load, see Jiajie Zhang and Donald A. Norman's "Representations in Distributed Cognitive Tasks" (1994) and Jiajie Zhang's "The Nature of External Representations in Problem Solving" (1997). Rather than restating note 2, the point here is its practical consequence: cognition is distributed across people, artifacts, environments, and time.

5. This is the central claim toward which notes 1, 2, and 4 converge. Hutchins's *Cognition in the Wild* (1995) provides the clearest system-level account, while Clark and Chalmers's "The Extended Mind" (1998) offers its philosophical articulation. Zhang and Patel's "Distributed Cognition, Representation, and Affordance" (2006) extends this framework by incorporating Gibson's affordance theory from The Ecological Approach to Visual Perception (1979), emphasizing how environments actively structure cognition. The point is not that individuals cease to matter, but that intelligence increasingly resides in the organization of interactions, representations, and workflows.

6. For the technical turning point in 2012, the essential citation is Alex Krizhevsky, Ilya Sutskever, and Geoffrey Hinton's "ImageNet Classification with Deep Convolutional Neural Networks" (2012). For the broader retrospective on why deep learning changed the field, see Yann LeCun, Yoshua Bengio, and Geoffrey Hinton's "Deep Learning" (2015). The historical point is that neural networks did not merely improve; they became scalable and practically dominant.

7. For the technical threshold behind large language models, see Tom Brown et al.'s "Language Models are Few-Shot Learners" (2020) and OpenAI's GPT-4 Technical Report (2023). For the interaction side—for how people increasingly work through iterative dialogue with AI rather than one-shot tool use—see Saleema Amershi et al.'s "Guidelines for Human-AI Interaction" (2019). The claim here is practical, not metaphysical: these systems now participate in the flow of reasoning, writing, and exploration.

8. This pattern is not unique to AI; it recurs whenever a general-purpose technology is added before institutions are redesigned around it. Paul David's "The Dynamo and the Computer" (1990) and Erik Brynjolfsson, Daniel Rock, and Chad Syverson's "The

Productivity J-Curve" (2021) both make that point clearly. The same dynamic helps explain why so much current AI activity looks energetic without yet being deeply transformational.

CHAPTER 1

The Illusion of Thinking Alone

We have been taught a simple story about intelligence:

It lives inside the individual.

We measure it through tests.
We cultivate it through education.
We reward it through credentials and status.

The smartest person is the one who knows the most, thinks the fastest, and solves problems independently.

This story is so familiar that it feels like a fact.

It is not.

The idea that thinking happens inside the individual mind is one of the most powerful—and most misleading— assumptions of the modern world.[1]

The Individual Mind as the Unit of Intelligence

Modern institutions are built on a shared premise:

That intelligence is an individual property.

Education is designed to transfer knowledge into individual minds.
Examinations measure what individuals can recall and apply.
Professions certify individuals based on their expertise.
Organizations reward individuals for their decisions and performance.

Even when we work in teams, we evaluate contribution at the level of the individual.

This assumption shapes everything:

> how we learn
>
> how we work
>
> how we lead
>
> how we organize

It is the invisible architecture beneath our systems.

Why This Model Worked

The model of individual intelligence did not emerge by accident.

It was well suited to a world in which:

> information was scarce
>
> access to knowledge was limited
>
> tools for externalizing cognition were constrained
>
> In that world:
>
> what you knew mattered
>
> what you could remember mattered
>
> what you could compute mentally mattered

The brain was the primary site of cognition because it had to be.

Under those conditions, it was reasonable to treat intelligence as something contained within individuals.

> The model was not wrong.
> It was contextually correct.

But its validity depended on constraints that no longer exist.

The Hidden Dependency

Even at its peak, the model of individual intelligence was incomplete.

Because no one has ever truly thought alone.[2]

Consider something as simple as reading.

When you read a book, where does the thinking happen?

In your mind?

On the page?

In the interaction between the two?

The answer is not obvious.

> The text holds structure, argument, and information.
> Your mind interprets, connects, and evaluates.

> Neither alone produces understanding.

> Thinking emerges from the interaction.

> This pattern appears everywhere.

> A scientist works with instruments, data, and models.
> A pilot relies on cockpit systems and checklists.
> A clinician coordinates with teams, records, and protocols.

> In each case:

> cognition is distributed
> reasoning is supported externally
> decisions emerge from systems[3]

The individual is central, but not sufficient.

The Myth of Self-Contained Thought

Despite this reality, we persist in treating cognition as self-contained.

We:

> isolate individuals in exams

> prohibit external aids in assessment

> celebrate independent thinking as the highest form of intelligence

These practices reinforce the belief that true cognition must occur inside the mind.

But they reflect an idealized version of thinking, not its actual form.

The mind has never been a closed system.

It has always depended on:

> external representations

> tools and artifacts

> interaction with others

What we call "individual intelligence" is, in many cases, system-supported cognition attributed to the individual.

Why the Illusion Persists

If cognition has always been distributed, why do we believe otherwise?

There are several reasons.

1. Visibility

Internal thought feels immediate and personal.
External contributions are often less visible.

We experience thinking as something happening "inside," even when it depends on outside structures.

2. Measurement

It is easier to measure individual performance than system-level cognition.

Exams, metrics, and evaluations focus on what individuals can do in isolation, because it is simpler to assess.

3. Institutional Design

Our systems are organized around individuals:

students

employees

leaders

Attributing cognition to individuals simplifies responsibility, accountability, and reward.

4. Cultural Narrative

We celebrate:

genius

individual achievement

independent insight

These narratives reinforce the idea that intelligence resides within individuals.

The Cost of the Illusion

As long as the constraints of the past held, this illusion was manageable.

But as cognition becomes increasingly distributed, the mismatch grows.

We begin to see the consequences:

> education that prioritizes memorization over understanding

> organizations that reward knowledge possession over judgment

> decision-making structures that ignore system-level dynamics

> governance models that cannot account for distributed responsibility

We are designing systems for a form of intelligence that no longer exists.

This creates inefficiency, misalignment, and missed opportunity.[4]

A Subtle Shift

The transformation we are now experiencing is not the creation of distributed cognition.

It is its amplification.

What has changed is not the existence of distributed cognition, but its scale, speed, and visibility.

With artificial intelligence:

> external systems become more capable

> interaction becomes more central

cognition becomes more obviously distributed

The illusion becomes harder to sustain.

From Individual to System

This leads to a fundamental shift.

From:

intelligence as an individual property

To:

intelligence as a system property[3]

This does not eliminate the role of the individual.

It redefines it.

The individual becomes:

a participant in a cognitive system

a contributor to distributed processes

a source of judgment and context

But no longer the sole locus of cognition.

A New Starting Point

If we accept that thinking is not confined to the mind, then many of our assumptions must be reconsidered.

What does it mean to learn?

What does it mean to know?

What does it mean to be intelligent?

These questions cannot be answered by looking at the individual alone.

They must be answered at the level of systems.

The Argument of This Book

This book begins from a simple but powerful premise:

Intelligence is not located in the individual.
It emerges from systems.

From this premise, a series of implications follows:

cognition is distributed

expertise is redefined

education must change

research must evolve

decision-making must be redesigned

governance must adapt

institutions must be rebuilt

Artificial intelligence is not the cause of this shift.

It is the catalyst that makes it unavoidable.[5]

Transition

To understand what is changing—and why—it is not enough to look at AI alone.

We must step back and examine how cognition has evolved over time.

How did thinking move beyond the brain?
What role did language, writing, and symbolic systems play?
And why does this trajectory make the emergence of AI not surprising, but inevitable?

The next chapter traces this long arc.

Not as a history of ideas, but as a history of how thinking itself has been constructed.

Endnotes

1. The assumption of individual intelligence is deeply built into modern schooling and testing. Lewis Terman's The Measurement of Intelligence (1916) is a canonical statement of that model, while Jean Lave's Cognition in Practice (1988) is one of the clearest correctives, because it shows how reasoning changes once we stop abstracting people away from the settings in which they actually think. The point here is not that individuals do not matter, but that institutions have long treated the individual as the complete unit of cognition.

2. For the distributed-cognition frame itself, the key texts are already in view from the Preface: Edwin Hutchins's Cognition in the Wild (1995), Donald Norman's Things That Make Us Smart (1993), and Jiajie Zhang and Donald Norman's "Representations in Distributed Cognitive Tasks" (1994). The practical lesson of that literature is exactly the one this chapter emphasizes: pilots, clinicians, and analysts do not reason alone; cognition is spread across people, tools, and representations.

3. Rather than opening a separate literature, this note sharpens note 2. Zhang and Norman's "Representations in Distributed Cognitive Tasks" (1994), Zhang's "The Nature of External Representations in Problem Solving" (1997) and James Hollan, Edwin Hutchins, and David Kirsh's "Distributed Cognition: Toward a New Foundation for Human-Computer Interaction Research" (2000) show how external representations, interfaces, and routines do part of the reasoning. Checklists, records, and dashboards are not merely aids to thought; they are components of thought.

4. The institutional mismatch point connects this chapter to the final note of the Preface. Paul David's "The Dynamo and the Computer" (1990) remains the classic account of why major technologies yield limited gains until organizations reorganize around them. Wanda Orlikowski's "Using Technology and Constituting Structures" (2000) makes the same point at the level

of everyday practice: when new tools are layered onto old routines, the appearance of change can outrun the reality of transformation.

5. For AI as catalyst rather than origin, the bridge runs between model capability and human-AI interaction. OpenAI's GPT-4 Technical Report (2023) captures the capability shift; Saleema Amershi et al.'s "Guidelines for Human-AI Interaction" (2019) and Fabrizio Dell'Acqua et al.'s "Navigating the Jagged Technological Frontier" (2026) show what happens when people begin to work with such systems in practice. The claim here is not that AI invents distributed cognition, but that it makes distributed cognition newly visible, scalable, and difficult for institutions to ignore.

CHAPTER 2

The Long Arc of Cognitive Evolution

Intelligence did not emerge fully formed in the human brain.
It was constructed—step by step—through the creation of systems outside it.[1]

For much of modern history, we have treated cognition as an internal property: something that happens inside the head, bounded by biology, measured by individual performance, and cultivated through personal learning. This assumption is so deeply embedded in education, science, and organizational design that it rarely gets questioned.

But it is wrong—or at least profoundly incomplete.

What we call "thinking" has never been confined to the brain. It has always depended on a scaffolding of external supports: symbols, tools, representations, and shared systems that extend cognition beyond the individual. These are not merely aids to thought.

They are part of thought itself.[2]

To understand artificial intelligence, we must first understand this longer arc: the evolution of cognition through the accumulation of cognitive artifacts.

Cognition Is Built, Not Given

The human brain is a remarkable organ, but it is not a self-sufficient thinking machine. It provides capacity—

perception, memory, pattern recognition—but the forms of cognition we associate with intelligence are not purely innate. They are constructed through interaction with external systems.

A child is not born knowing mathematics, logic, or even language in its full expressive form. These are acquired—not just through internal development, but through immersion in symbolic environments that pre-exist the individual. Language is learned through participation in a linguistic community. Writing is learned through exposure to a system of external representation. Mathematical reasoning emerges only after engagement with formal symbolic structures.

This is the first critical insight:

> Human cognition is not simply developed within the brain.
> It is assembled through systems that extend beyond it.[2]

The implication is profound. If cognition depends on external structures, then changes to those structures change cognition itself.

The First Extensions: Thinking Together

Long before writing or mathematics, early humans extended cognition through the body and through social coordination.

Gesture, imitation, and shared attention allowed groups to coordinate action. Knowledge was not stored in individuals alone, but in patterns of interaction—rituals, practices, and embodied behaviors passed from one person to another.

With the emergence of language, this extension took a decisive leap.

Language did more than enable communication. It created shared cognitive space. Thoughts could be expressed, negotiated, refined, and transmitted across individuals. Meaning no longer resided solely within a single mind, but in the interaction between minds.[3]

Language did not just allow us to speak.
It allowed us to think together.

This was the first large-scale redistribution of cognition—from the individual to the collective.

Writing: Memory Outside the Brain

The invention of writing marked another transformation. For the first time, memory could be externalized in a stable, durable form.[4]

Before writing, knowledge had to be remembered, rehearsed, and transmitted orally. Memory was fragile, limited, and subject to distortion. Writing changed that.

It created a new kind of cognitive system in which information could persist independently of any individual. Knowledge could accumulate across generations, be revisited, revised, and recombined.

Writing is not a tool for recording thought.
It is a system that makes new forms of thought possible.

Complex arguments, long chains of reasoning, and large bodies of knowledge become feasible only when memory is no longer constrained by the limits of the brain. The external page becomes part of the cognitive process.

When we read, we are not simply retrieving information. We are interacting with an external memory system that shapes how we think.

Mathematics and Logic: Reasoning Outside the Brain

If writing externalized memory, mathematics and logic externalized reasoning.[5]

Mathematical notation and formal logic provide structured systems in which relationships can be represented, manipulated, and evaluated with precision. They reduce ambiguity, enforce consistency, and allow reasoning to be carried out step by step in a shared symbolic space.

These systems do not merely describe reasoning. They enable it.

A complex proof, for example, is not held entirely in the mind. It unfolds across symbols, diagrams, and intermediate steps that exist outside the brain. The thinker interacts with these representations, using them to guide and constrain thought.

> Mathematics is not just a language of description.
> It is a machinery of thinking.

Once again, cognition expands—not by increasing the brain's capacity, but by offloading and structuring processes in external systems.

From Individuals to Systems

Across these developments, a pattern emerges:

> Language distributes cognition across people
> Writing distributes cognition across time
> Mathematics distributes cognition across symbolic systems

Each step moves cognition further away from the individual brain and deeper into networks of interaction between people and artifacts.

At this point, the idea that intelligence resides solely within individuals becomes difficult to sustain.

Cognition is increasingly a property of systems:[3]

groups of people

shared representations

structured environments

> We do not think alone.
> We think with systems.

Digital Systems: Distributed Information

The rise of digital technology accelerated this process dramatically.

Computers and networks expanded the scale and speed of distributed cognition. Information could now be stored, retrieved, and transmitted globally, in real time. Databases replaced archives. Search replaced recall. Networks replaced isolated knowledge.

The constraint of access—once a major limitation on cognition—began to dissolve.

But even in this phase, most cognitive artifacts remained fundamentally passive. They stored information, transmitted it, or organized it. They did not actively participate in thinking.

The human remained the primary agent of cognition.

The Threshold

Across all these stages, cognitive artifacts shared a common characteristic:

> They extended cognition, but they did not generate it.

They supported thinking.
They did not think.

This distinction is crucial, because it is precisely what artificial intelligence changes.[6]

AI: A New Class of Cognitive Artifact

Artificial intelligence represents a fundamental shift in the nature of cognitive artifacts.

Unlike language, writing, or mathematics, AI does not merely store, transmit, or structure information. It generates. It evaluates. It interacts. It adapts.

It participates in cognition.[6]

For the first time in history, the cognitive artifact is not passive. It is active.

It does not just hold thought.
It contributes to it.

This marks a discontinuity in the evolution of cognition.

If writing allowed us to remember beyond ourselves, and mathematics allowed us to reason beyond ourselves, AI allows us to engage with a system that produces reasoning alongside us.

The boundary between the thinker and the tool begins to blur.

Cognition Relocates

With this shift, the location of cognition changes again.

No longer confined to:

the brain

the page

or the symbolic system

Cognition now emerges through interaction between humans and intelligent artifacts.

This leads to a second critical insight:

Intelligence is no longer a property of individuals.
It is a property of human–AI systems.

The unit of cognition has changed.

We are not simply using tools more effectively.
We are participating in systems that think.

The Implication

This long arc—from gesture to language, from writing to mathematics, from digital systems to AI—reveals a consistent pattern:

Human intelligence evolves by moving beyond the human.

Each major advance did not enhance cognition within the brain.
It reconfigured cognition across systems.

Artificial intelligence is the latest—and most powerful—step in this trajectory. It is not an anomaly; it is the logical endpoint of a long process in which cognition is progressively externalized, structured, and scaled beyond the individual.

But it is also different in kind.

Previous cognitive artifacts extended what we could do.
AI changes what cognition is.

A New Question

If cognition is no longer located within individuals, then many of our foundational assumptions begin to unravel.

What does it mean to learn, if knowledge is distributed?

What does it mean to be an expert, if reasoning is shared?

What does it mean to make decisions, if cognition is collaborative?

And perhaps most importantly:

> What does it mean to design institutions in a world where no one thinks alone?

> These are not questions about technology.
> They are questions about the nature of intelligence itself.

Transition

To answer them, we must move beyond the history of cognitive artifacts and confront what AI is actually doing to cognition in the present.

> Not making it faster.
> Not making it easier.

> But reorganizing it.

> The next chapter examines this transformation directly:

How AI shifts thinking from an internal process to a system-level phenomenon—and why this change is already redefining work, expertise, and decision-making.

Endnotes

1. Rather than rehearse the distributed-cognition case already laid out in the Preface and Chapter 1, the key historical source here is Merlin Donald's Origins of the Modern Mind (1991). Donald's

argument is that human cognitive evolution is driven not only by biology but by successive external symbol systems. That is the deeper backdrop for this chapter's claim that cognition is built, not simply given.

2. For the tighter claim that external representations are part of thinking, the best fit remains Edwin Hutchins's Cognition in the Wild (1995), Jiajie Zhang and Donald Norman's 'Representations in Distributed Cognitive Tasks' (1994), and Jiajie Zhang's 'The Nature of External Representations in Problem Solving' (1997). These works make clear that symbolic environments do not merely support cognition after the fact; they reorganize what kind of reasoning is possible in the first place.

3. The point about language as shared cognitive space is most closely associated with Lev Vygotsky's Mind in Society (1978). Vygotsky's central idea is that higher psychological functions are socially mediated before they are internalized. In that sense, language is not only a vehicle for expressing thought; it is one of the environments in which thought is formed.

4. On writing as memory outside the brain, Walter Ong's Orality and Literacy (1982) and Jack Goody's The Logic of Writing and the Organization of Society (1986) are still the cleanest sources. Both show, in different ways, that writing changes not just storage but the structure of argument, administration, and reflection. The significance here is not archival convenience; it is cognitive reorganization.

5. For the claim that mathematics and logic externalize reasoning, Allen Newell and Herbert Simon's Human Problem Solving (1972) and Herbert Simon's The Sciences of the Artificial (1969) remain foundational. Their shared lesson is that formal symbolic systems change the search space of reasoning itself. This is also where Zhang's 1997 analysis of external representations remains useful: the representation is part of the problem-solving machinery.

6. AI matters here because it becomes an active, not merely passive, cognitive artifact. Tom Brown et al.'s 'Language Models are Few-Shot Learners' (2020) and OpenAI's GPT-4 Technical Report (2023) mark the capability shift, while Daniil Boiko, Robert MacKnight, Ben Kline, and Gabe Gomes's 'Autonomous Chemical

Research with Large Language Models' (2023) shows what that shift looks like when a model participates in experimental reasoning rather than simply storing or transmitting information.

CHAPTER 3

When Thinking Leaves the Mind

We tend to describe artificial intelligence as making us faster, more efficient, more productive.

> This is a comforting story.
> It is also the wrong one.

> AI is not making thinking faster.
> It is moving thinking.

> AI is not simply accelerating thinking.
> It is reorganizing where thinking happens.[1]

The shift is subtle at first. A clinician consults a model before forming a diagnosis. A researcher uses AI to generate hypotheses before refining them. A student drafts an argument through interaction with a system rather than composing it alone.

In each case, the human is still present. The output still appears human. The activity still feels like thinking.

But the location of cognition has changed.

> Thinking is no longer an internal process occasionally supported by tools.
> It is an external process in which the human participates.

This is the beginning of a new cognitive regime.

From Support to Participation

For centuries, cognitive artifacts have supported thinking without participating in it.

A book stores knowledge.
A notebook extends memory.
A formula structures reasoning.

But none of these artifacts generates ideas on its own. They require a human to activate them, interpret them, and carry the cognitive process forward.

Artificial intelligence breaks this pattern.

It does not wait.
It responds.

It does not merely store representations.
It produces them.

It does not only structure reasoning.
It performs it.

When you interact with an AI system, you are no longer working with a passive medium. You are engaging with a system that actively contributes to the cognitive process—proposing interpretations, generating alternatives, surfacing patterns, and even anticipating your intent.[2]

For the first time, the artifact is no longer a container of thought.
It is a participant in thought.

This is not an incremental improvement. It is a categorical shift.

Language Becomes Computation

The most profound transformation is happening through language.

For most of human history, language has been the medium of thought. It allows us to represent ideas, communicate them, and refine them through interaction. But language itself was not executable. It described thinking; it did not perform it.

That boundary has now dissolved.

Large language models transform language into a computational substrate. Words are no longer just symbols for communication; they are inputs to systems that generate reasoning, structure arguments, and simulate understanding.[3]

> You do not simply write with AI.
> You think through it.

> A prompt becomes a cognitive act.
> A response becomes part of the reasoning process.

The interaction itself becomes the site of cognition.

> Language is no longer just a medium of thought.
> It is now a platform for thinking.

This shift is easy to underestimate because it feels natural. We are accustomed to thinking in language. But what has changed is not the medium—it is the agency within the medium.

Language now contains a second mind.

The Externalization of Cognitive Processes

As AI systems become more capable, core cognitive functions begin to move outside the individual:

Idea generation is initiated by models

Explanation is co-produced through dialogue

Synthesis emerges from iterative interaction

Evaluation becomes a shared process

This does not mean that humans stop thinking. It means that thinking becomes distributed across the interaction between human and system.[4]

Consider writing.

Traditionally, writing was an internal process externalized onto the page. The author generated ideas, structured them mentally, and then encoded them in text.

Now, writing often begins externally:

an AI drafts an outline

proposes arguments

generates variations

The human responds, edits, selects, and redirects.

Where is the thinking happening?

Not solely in the human.
Not solely in the machine.
But in the loop between them.

Cognition has moved from the individual to the interaction.

The Loop Becomes the Unit

This interaction—this continuous exchange between human and AI—is the new unit of cognition.

It is not a tool-use relationship.
It is a coupled system.

The human provides:

goals

context

judgment

constraints

The AI provides:

generation

pattern recognition

expansion

simulation

Neither alone constitutes the full cognitive process.

The unit of intelligence is no longer the individual.
It is the interaction.

Intelligence emerges from the loop.[4]

This has two important consequences.

First, the performance of the system cannot be understood by evaluating either component in isolation.

Second, cognitive skill shifts from internal processing to interaction design.

The question is no longer:

"How well can you think?"

It becomes:

"How well can you structure a system that thinks?"

From Recall to Recognition, From Creation to Evaluation

As cognition relocates, the nature of cognitive work changes.

Tasks that once defined intelligence begin to diminish in importance:

Memorizing information

Generating first drafts

Performing routine analysis

These functions are increasingly handled by AI systems.

In their place, new forms of cognitive work emerge:

Recognition over recall — identifying relevant patterns rather than retrieving stored knowledge

Verification over generation — assessing correctness rather than producing content

Evaluation over creation — selecting, refining, and judging among alternatives

This is not a reduction of human intelligence. It is a redistribution.

Humans move toward:

framing problems

setting direction

exercising judgment under uncertainty

AI systems handle:

expansion of possibilities

rapid generation

probabilistic reasoning at scale

Intelligence shifts from producing answers to governing answers.

The Illusion of Individual Performance

Despite this transformation, we continue to evaluate intelligence as if it resides within individuals.

> We test students on what they can recall.
> We assess professionals on what they can produce independently.
> We benchmark systems on isolated performance metrics.

These measures increasingly fail to capture where cognition actually occurs. Empirical studies already show that performance degrades when human–AI interaction is poorly structured, even when the underlying models perform well in isolation.[5]

The relevant unit is not the individual, but the system they are part of.

We are measuring the wrong thing.

Intelligence is no longer an individual attribute.[5]

Failure as a System Property

When outcomes degrade—when errors occur—we often attribute failure to the model or the user.

But in distributed cognition, failure is a system property.

A breakdown can occur because:

> mis-specified prompts
>
> misinterpretation of intent
>
> over-trust or under-trust
>
> poor interaction structure

The result is not a failure of human intelligence or artificial intelligence alone, but of the coordination between them.

The critical design problem is no longer model accuracy. It is alignment within the cognitive system.

The New Cognitive Boundary

In earlier eras, the boundary of cognition was clear: it ended at the skin.

Even as artifacts extended thinking, the human remained the central locus of cognition.

That boundary has now dissolved.

Cognition extends into:

interfaces

prompts

models

data environments

The "mind" is no longer a contained entity. It is a dynamic system spanning biological and artificial components.

The brain is no longer the boundary of intelligence.

A New Form of Thinking

What emerges from this transformation is not enhanced individual cognition, but a new form of thinking altogether.

It is:

interactive rather than internal

iterative rather than linear

distributed rather than localized

It unfolds through dialogue, not monologue.

The thinker is no longer a solitary agent, but a participant in a cognitive system.

This does not diminish human agency. It changes its role.

From:

> generator
>
> holder of knowledge
>
> sole decision-maker

To:

> orchestrator
>
> evaluator
>
> architect of cognitive systems

Transition

If thinking is no longer an individual act, then expertise—long defined by individual knowledge and skill—must also change.

> What does it mean to be an expert in a world where cognition is distributed?
> What does it mean to know, when knowledge is externalized and dynamically generated?

The next chapter examines the collapse and reconstruction of expertise in the age of AI—and why the most valuable cognitive skill is no longer knowing more, but judging better.

Endnotes

1. The base claim that cognition can relocate into an interaction loop has already been established earlier; the sharper source here is Hollan, Hutchins, and Kirsh's 'Distributed Cognition: Toward a

New Foundation for Human-Computer Interaction Research' (2000). Read alongside Hutchins's Cognition in the Wild (1995), it explains why the relevant unit is no longer the isolated thinker but the coupled system.

2. For AI as a participant rather than a passive aid, Brown et al.'s 'Language Models are Few-Shot Learners' (2020) and OpenAI's GPT-4 Technical Report (2023) show the capability shift, while Boiko et al.'s 'Autonomous Chemical Research with Large Language Models' (2023) shows models planning and executing research steps with tools. The point is not that the machine 'has a mind' in a human sense; it is that the artifact now contributes non-trivially to the flow of cognition.

3. The claim that language has become computational is best anchored in Brown et al. (2020) and GPT-4 (2023). Once natural language becomes a control surface for a generative system, prompts stop being mere instructions and become operators over a reasoning substrate. That is why prompt-response interaction feels less like querying a database and more like thinking in dialogue.

4. For the idea that the loop itself becomes the unit, Saleema Amershi et al.'s 'Guidelines for Human-AI Interaction' (2019) is still the most useful design reference, and Daniel McDuff et al.'s 'Towards Accurate Differential Diagnosis with Large Language Models' (2025) makes the point empirically in a high-stakes domain. Both suggest that outcomes depend on how humans and AI are coupled, not on model quality alone.

5. The same lesson appears in jagged-frontier work. Fabrizio Dell'Acqua et al.'s 'Navigating the Jagged Technological Frontier' (2026) shows that AI can improve performance on some tasks and degrade it on others within the same workflow. That is exactly why this chapter treats failure as a system property rather than as a defect attributable only to the human or the model.

CHAPTER 4

The Collapse and Reconstruction of Expertise

For centuries, expertise has been defined by possession.

To be an expert was to know more—to hold within one's mind a large body of knowledge, refined through training, experience, and repetition. Expertise meant recall, mastery, and the ability to generate correct answers under uncertainty.

This definition is now breaking.

Not because expertise is no longer valuable, but because the conditions that made it scarce have changed.

When knowledge becomes abundant, possession is no longer the basis of expertise.

> Artificial intelligence does not eliminate expertise. It destabilizes its foundation.[1]

The Old Model: Expertise as Internal Knowledge

Traditional expertise rests on three assumptions:

Knowledge is scarce

Knowledge must be internalized

Performance depends on recall and synthesis within the individual

This model shaped:

education (memorization and mastery)

professions (credentialing and specialization)

organizations (hierarchies based on knowledge gradients)

The expert was the one who knew.

This worked in a world where:

information was limited

access was constrained

cognitive labor was expensive

Under those conditions, internal knowledge was both necessary and sufficient.

The Break: Knowledge Becomes External and Dynamic

AI systems fundamentally alter these assumptions.

Knowledge is no longer scarce—it is instantly accessible

It no longer needs to be fully internalized—it can be externalized

It is no longer static—it is dynamically generated

A clinician no longer relies solely on memory for differential diagnosis.
A lawyer does not need to recall every precedent.
A researcher does not begin from a blank page.

In each case, AI expands the accessible cognitive space far beyond what any individual can contain.[2]

This creates a structural shift:

The value of knowing decreases as the cost of accessing knowledge approaches zero.

But this does not make expertise obsolete.
It changes what expertise is.

This does not mean expertise disappears.
It means knowledge is no longer the constraint—
judgment is.

From Knowing to Judging

When answers can be generated instantly, the critical
question is no longer:

"Do you know?"

It becomes:

"Can you judge?"

Judgment includes:

assessing correctness

recognizing relevance

identifying limitations

understanding context

making decisions under uncertainty

AI can produce plausible answers.
It cannot fully determine when those answers are
appropriate, sufficient, or trustworthy within a specific
context.[3]

Expertise shifts from producing answers to governing
answers.

This is not a lesser role.
It is a more demanding one.

The Paradox of Capability

As AI systems become more capable, the demand for expertise increases—not decreases.

This appears counterintuitive.

If AI can generate high-quality outputs, why do we need experts?

Because:

More outputs create more choices

More choices increase complexity

Increased complexity requires better judgment

A less capable system produces fewer options.
A more capable system produces many plausible options.

Selecting among them becomes the central task.

Abundance of answers creates scarcity of judgment.

The Illusion of De-skilling

There is a common concern that AI will "de-skill" professionals—that reliance on external systems will erode internal capability.

This concern is not entirely unfounded, but it is incomplete.

What appears as de-skilling is often re-skilling.

Skills tied to:

recall

routine generation

repetitive analysis

become less central.

Skills tied to:

interpretation

evaluation

integration

decision-making

become more central.

The surface of expertise changes, but its depth increases.

Expertise does not disappear.
It migrates.

Distributed Expertise

In the traditional model, expertise is located within individuals.

In the emerging model, expertise is distributed across:

humans

AI systems

data environments

workflows

No single component holds complete expertise.

It emerges from the interaction between them.[4]

Consider a decision-making scenario:

The human frames the problem

The AI generates possibilities

The human evaluates and constrains

The system iterates

Where is the expertise?

Not in any single element, but in the system as a whole.

Expertise becomes a property of the human–AI system.

The New Expert

If expertise is no longer defined by possession of knowledge, what defines the expert?

The new expert is characterized by the ability to:

Frame problems precisely

defining the right question becomes more important than knowing the answer

Structure interactions with AI systems

guiding, constraining, and iterating effectively

Evaluate outputs critically

distinguishing signal from noise, correctness from plausibility

Integrate across domains

combining insights from multiple sources and perspectives

Exercise judgment under uncertainty

making decisions when answers are incomplete or ambiguous

This is a fundamentally different cognitive profile.

The expert is no longer the one who knows the most. It is the one who navigates the system best.

The Risk: False Expertise

As AI lowers the barrier to generating sophisticated outputs, it also lowers the barrier to appearing competent.

Individuals can produce:

convincing arguments

technical explanations

domain-specific analyses

without deep understanding.

This creates a new risk:

The proliferation of plausible but shallow expertise.

When outputs look correct, but are not deeply grounded, evaluation becomes more difficult.

In this environment:

surface fluency can mask lack of understanding

confidence can outpace competence

errors can propagate more quickly

This increases the importance of true expertise—not as production, but as evaluation.

Credentialing Without Cognition

Our current systems of education and professional certification are still based on the old model of expertise.

They test:

recall

independent performance

knowledge possession

But these measures increasingly fail to capture the skills that matter in a distributed cognitive system.[5]

A person may perform well on traditional assessments yet struggle to:

work effectively with AI

evaluate generated outputs

make sound decisions in complex systems

Conversely, someone skilled in orchestrating human–AI interaction may outperform traditional experts in real-world settings.

We are certifying the wrong capabilities.

This creates a growing gap between:

formal expertise

functional expertise

Reconstructing Expertise

To align with the new reality, expertise must be redefined.

Not as:

what you know

or what you can produce alone

But as:

how you think within a system

how you coordinate cognition across human and machine

how you exercise judgment in a distributed environment

This requires changes at multiple levels:

Education must shift from knowledge acquisition to cognitive orchestration

Professional training must emphasize evaluation and decision-making

Organizations must redesign roles around system-level performance

Expertise becomes less about accumulation and more about navigation.

A New Hierarchy

In the old model, hierarchy followed knowledge gradients:

those who knew more held authority

In the new model, hierarchy follows judgment gradients:

those who can evaluate and decide effectively hold authority

This is a quieter but more profound shift.

Knowledge can be accessed by many.
Judgment remains unevenly distributed.

The future will not belong to those who know the most.
It will belong to those who decide the best.

Transition

If expertise is no longer an individual possession, then the structures built around it—education systems, professional hierarchies, organizational designs—must also change.

Yet most institutions are not redesigning themselves.

They are adapting at the surface while preserving their underlying assumptions.

They are not transforming.
They are performing transformation.

The next chapter examines this phenomenon—why so many organizations appear to be innovating while remaining fundamentally unchanged, and why this pattern is not intentional, but structural.

This is the trap of innovation theater.

Endnotes

1. The older view of expertise as possession is well captured by The Cambridge Handbook of Expertise and Expert Performance (2006), edited by K. Anders Ericsson et al., while Harry Collins and Robert Evans's Rethinking Expertise (2007) helps explain why expertise is never just a pile of facts. The move this chapter makes is not away from expertise but away from defining it by stored knowledge alone.

2. On expanded cognitive access, the important bridge is between model capability and real use. OpenAI's GPT-4 Technical Report (2023) shows the widening accessible problem space; Daniel McDuff et al.'s 'Towards Accurate Differential Diagnosis with Large Language Models' (2025) shows what happens when that expanded space is placed inside a professional workflow. The expert no longer starts from a blank page or bare memory.

3. The limit of AI is not fluency but judgment. Emily M. Bender, Timnit Gebru, Angelina McMillan-Major, and Shmargaret Shmitchell's 'On the Dangers of Stochastic Parrots' (2021) remains the sharpest warning about plausible but ungrounded language, and GPT-4's own technical report (2023) explicitly documents hallucination and reliability limits. That is why this chapter treats judgment as the scarce capability.

4. The distributed-expertise point does not require a new literature beyond what has already been established; it requires a sharper use of it. Hutchins's Cognition in the Wild (1995), Zhang and Norman's 'Representations in Distributed Cognitive Tasks' (1994), and Zhang's 'The Nature of External Representations in

Problem Solving' (1997) together show how competence can reside in coordinated systems rather than in a single head.

5. The assessment problem is now visible in AI settings. Amershi et al. (2019) shows that human-AI performance depends on interaction design, Daniel McDuff et al. (2025) shows that clinicians using a well-designed conversational aid can outperform standard-resource use, and Dell'Acqua et al. (2026) shows that naive use can still backfire. Formal expertise and functional expertise are beginning to separate.

CHAPTER 5

The Trap of Innovation Theater

Every major technological shift produces two kinds of change:

Visible activity

Structural transformation

They are not the same.

In the age of artificial intelligence, most organizations are very active. They are launching initiatives, forming committees, piloting tools, and announcing strategies. From the outside, it appears that transformation is underway.

But in many cases, nothing fundamental has changed.

The appearance of innovation is not the same as the reality of transformation.

This gap—between what organizations do and what they become—is what we call innovation theater.[1]

Innovation theater is what happens when activity substitutes for redesign.

Not Designed, But Emergent

Innovation theater is rarely intentional.

Leaders are not trying to mislead. Institutions are not deliberately resisting change. In most cases, efforts to adopt AI are thoughtful, well-motivated, and sincere.

Yet the outcome often looks the same:

   new AI centers layered onto existing structures

   pilot projects disconnected from core workflows

   task forces without authority or execution pathways

   new roles without budget or decision rights

These efforts create motion without direction.

   Innovation theater is not designed.
   It emerges when new technologies are layered onto old
   architectures.[2]

The Structural Mismatch

At the core of innovation theater is a mismatch between two
systems moving at very different speeds:

   Technology evolves exponentially

   Institutions evolve linearly

AI capabilities are advancing rapidly—models improve in
months, not decades. But institutions are governed by
processes that assume stability:

   annual budget cycles

   multi-year strategic plans

   accreditation and regulatory timelines

   hierarchical decision-making structures

These systems were designed for the industrial era, where
change was gradual and predictable.

They are poorly suited for a world where capabilities
compound quickly and continuously.

Institutions are trying to absorb exponential change with linear mechanisms.

The result is not failure.
It is stagnation disguised as progress.[3]

Surface Change vs. Architectural Change

To understand innovation theater, we must distinguish between two types of change:

Surface Change

adding new tools

creating new programs

launching pilots

issuing policies

Architectural Change

redesigning workflows

restructuring decision-making

reallocating authority and resources

redefining roles and incentives

Most organizations focus on surface change because it is:

faster

safer

more visible

easier to communicate

But surface change leaves the underlying system intact.

When architecture does not change, behavior does not change.

AI becomes an add-on rather than a foundation.[4]

The Parallel System Problem

A common pattern in innovation theater is the creation of parallel systems.

An organization introduces:

> an AI innovation lab

> a digital transformation office

> a pilot program for new workflows

But these operate alongside existing systems rather than replacing them.

> The legacy system continues to function as before.
> The new system operates in isolation.

> Over time:

> the pilot remains a pilot

> the innovation lab remains peripheral

> the core system remains unchanged

Innovation is contained rather than integrated.

This creates the illusion of progress without the reality of transformation.

Why This Pattern Persists

Innovation theater persists not because organizations misunderstand AI, but because of structural constraints.

1. Risk Aversion

Fundamental redesign is disruptive. It threatens existing roles, processes, and power structures.

Surface change allows organizations to signal progress without incurring these risks.

2. Incentive Misalignment

Leaders are rewarded for:

launching initiatives

demonstrating activity

producing short-term results

They are rarely rewarded for:

dismantling legacy systems

redesigning institutional architecture

pursuing long-term transformation

3. Organizational Inertia

Institutions accumulate:

policies

procedures

cultural norms

These create resistance to change—not through opposition, but through friction.

4. Cognitive Framing

Perhaps most importantly, organizations still conceptualize AI as a tool rather than as infrastructure.

If AI is seen as a tool, it is added.
If AI is understood as infrastructure, the system must be redesigned.

The problem is not execution.
It is how the problem is framed.

The Illusion of Progress

Innovation theater is effective because it produces visible signals of change:

press releases

new titles

dashboards

pilot results

These signals are convincing—to stakeholders, to leadership, and even to those inside the organization.

But they do not necessarily reflect underlying transformation.

An institution can appear highly innovative while remaining structurally unchanged.

Activity creates the illusion of progress.
Architecture determines whether progress is real.[5]

Innovation theater is what happens when activity substitutes for redesign.

The Cost of Staying Superficial

In the early stages of technological change, innovation theater can persist without immediate consequences.

Organizations can:

experiment

explore

delay deeper decisions

But as the technology matures, the cost of superficial adoption increases.

competitors redesign workflows and gain efficiency

new entrants build AI-native systems from the ground up

performance gaps widen

At that point, incremental adaptation is no longer sufficient.

Systems that are not redesigned become uncompetitive.

The risk is not that organizations will fail to adopt AI.

It is that they will adopt it in ways that do not matter.

From Theater to Transformation

Escaping innovation theater requires a shift in perspective.

From:

"How do we use AI?"

To:

"How must we redesign the system if AI is part of it?"

This leads to a different set of questions:

How do workflows change when cognition is distributed?

How do roles change when expertise is shared with AI?

How do decisions change when reasoning is externalized?

How do we allocate authority in human–AI systems?

These are architectural questions, not operational ones.

They cannot be answered through pilots alone.

The First Principle of Real Transformation

At its core, the transition out of innovation theater requires a single principle:

> AI must be treated as infrastructure, not as a tool.[6]

> Infrastructure is not optional.
> It is not peripheral.
> It shapes everything built on top of it.

> When electricity became infrastructure, factories were redesigned.
> When the internet became infrastructure, organizations were restructured.

> AI will have the same effect.

> But only if it is recognized as such.

A Diagnostic Test

An organization can ask a simple question to assess whether it is engaged in innovation theater:

> If we removed all our AI initiatives tomorrow, would our core operations fundamentally change?

If the answer is no, then AI has not yet been integrated at the architectural level.

It remains a layer, not a foundation.

Innovation theater is what happens when activity substitutes for redesign.

Transition

If institutions must move beyond innovation theater, the next question becomes unavoidable:

What does it mean to redesign a system for a world in which cognition is distributed and AI is infrastructure?

But before institutions can be redesigned, we must first confront a deeper shift:

AI is not simply a tool being added to existing systems. It is becoming a medium through which cognition itself occurs.

This distinction is decisive.

As long as AI is treated as a tool, organizations will continue to layer it onto existing architectures—and remain trapped in innovation theater.

Only when AI is understood as a medium does redesign become inevitable.

The next chapter examines this transition—from tool to medium—and why it fundamentally changes how organizations think, learn, and decide.

Endnotes

1. 'Innovation theater' is newer as a phrase than as a phenomenon. The underlying pattern is already visible in Paul David's 'The Dynamo and the Computer' (1990): organizations install a general-purpose technology before they reorganize around it. Wanda Orlikowski's 'Using Technology and Constituting Structures' (2000) makes the same point at the level of everyday practice.

2. The failure mode of layering new technology onto inherited workflows is one of the central findings of digital transformation research. Orlikowski (2000) explains why tools do not determine practice on their own, and Erik Brynjolfsson, Daniel Rock, and Chad Syverson's 'The Productivity J-Curve' (2021) shows why

measurable gains often lag until complements, processes, and organizational intangibles catch up.

3. The 'exponential technology, linear institution' contrast is a compressed way of describing a broader structural mismatch. Timothy Bresnahan and Manuel Trajtenberg's 'General Purpose Technologies "Engines of Growth"?' (1995) and Brynjolfsson, Rock, and Syverson (2021) both suggest that once a technology is infrastructural, the limiting factor becomes institutional adaptation, not invention alone.

4. This note simply sharpens note 1. David (1990) and Orlikowski (2000) both show that adopting a new technical layer is not the same thing as redesigning a system around it. That is why pilots and labs so often create motion without architectural change.

5. On the illusion of progress, Marshall Meyer and Vipin Gupta's 'The Performance Paradox' (1994) is still useful: visible metrics can drift away from underlying performance. Read together with Brynjolfsson, Rock, and Syverson (2021), it helps explain why dashboards and announcements can rise even while institutional capability remains flat.

6. The infrastructure argument rests on the classic general-purpose-technology literature. Bresnahan and Trajtenberg (1995), David (1990), and Brynjolfsson, Rock, and Syverson (2021) all point to the same conclusion: value does not come from bolt-on use, but from redesigning the system that sits on top of the technology.

CHAPTER 6

From Tool to Medium

We tend to describe artificial intelligence as making us faster, more efficient, more productive.

> This is a comforting story.
> It is also the wrong one.

> AI is not making thinking faster.
> It is moving thinking.

> AI is not simply accelerating thinking.
> It is reorganizing where thinking happens.[1]

Most organizations believe they are adopting AI.

In reality, they are making a deeper choice—often without realizing it:

> What kind of cognition will our system have?

> Will AI remain a tool—improving efficiency at the edges?

> Or will it become something else entirely—a medium through which thinking itself occurs?

> AI is not a tool you use.
> It is a medium you think within.

> This distinction is not technical.
> It is cognitive.
> And ultimately, it is institutional.

The Hidden Question Beneath AI Adoption

When organizations talk about AI maturity, they usually mean:

how many models are deployed

how much data is integrated

how advanced their platforms are

These are important.

But they miss the deeper question:

What kind of thing is AI becoming in our system?

If AI is treated as a tool:

it optimizes tasks

it accelerates workflows

it improves efficiency

If AI becomes a medium:

it reshapes how problems are framed

it transforms how knowledge is created

it reorganizes how decisions are made

The difference is not degree.
It is kind.[2]

The Long Arc: Cognition Changes with Representation

Across history, cognition has changed when representational systems changed.

In oral cultures, knowledge lived in people and practices

Writing externalized memory and enabled reflection

Print stabilized knowledge and scaled verification

Digital networks made information shareable and recombinable

Each shift did not make people "more intelligent."

It changed:

what could be represented

how it could be manipulated

how it could be shared

New representations create new forms of thinking.[3]

Artificial intelligence continues this trajectory—but introduces something fundamentally new.

It does not just store or transmit representations.

It transforms them.

From Information to Representational Work

The internet made information widely available.

AI makes representational work widely available.

It can:

summarize

translate

generate

simulate

reframe

Work that once required expertise and time can now be invoked on demand.

This is the turning point.

The internet made knowledge shareable.
AI makes thinking shareable.[4]

And that is the bridge from tool to medium.

The Four Levels of AI Maturity

To understand this transition, we need a different way to think about AI maturity.

Not as technology adoption.

But as changes in representational ecology—the system of representations through which an organization thinks.[3]

AI maturity is not defined by how much AI you have.

It is defined by how cognition is organized.

Level 1 — AI as Specialty Craft

At this level, AI capability is concentrated in specialists.

Data scientists build models

Outputs are delivered as reports or dashboards

Most people consume results passively

AI appears as:

> predictions

> classifications

> forecasts

> The cognitive workflow remains unchanged.

> Experts think.
> Others receive.

AI improves accuracy and efficiency, but does not change how the organization thinks.

Level 2 — AI as Literacy

AI literacy becomes as fundamental as reading and writing were in the industrial age.

At this level, AI use becomes widespread.

People begin to:

> interact directly with AI systems

> understand limitations and uncertainty

> critique outputs

> AI outputs become a new kind of "text":

> readable

> interpretable

> debatable

> Norms begin to shift:

> evidence is evaluated differently

> uncertainty becomes explicit

> interpretation becomes a shared activity

> But the system is still fragmented.

> People use AI.
> But the organization does not yet think through AI.

Level 3 — AI-Native Architecture

At this level, AI becomes embedded in workflows.

Processes are redesigned

Feedback loops are continuous

data flows across systems

The unit of performance shifts:

From:

individual work

To:

human–AI systems

Information is no longer static.

It is:

generated

transformed

propagated

Cognition becomes distributed in practice.[5]

The system begins to think.

Level 4 — AI as Cognitive Medium

This is the threshold.

AI is no longer a tool within the system.

It becomes part of the system's epistemic infrastructure—how it knows.

Knowledge itself changes form.

It becomes:

dynamic rather than static

continuously updated rather than periodically revised

interactive rather than fixed

AI participates in:

generating hypotheses

exploring alternatives

refining explanations

supporting epistemic search

Thinking happens within the medium.

A tool supports cognition.
A medium hosts it.

The Diagnostic Map: Where Are You Now?

Not all AI initiatives sit at the same level.

They can be mapped across two dimensions:

Degree of integration (localized → systemic) — how deeply AI is embedded in core workflows and decision processes.

Nature of impact (operational → epistemic) — whether AI changes task execution or fundamentally reshapes how knowledge is produced and decisions are made.

Taken together, these dimensions reveal four distinct regimes of AI use:

Most organizations cluster in the upper-left quadrant: localized, operational use—tools and pilots that create visible activity but limited structural change.

Progress requires movement along two axes:

From localized → systemic (integration into core workflows)

From operational → epistemic (shifting from doing work to reshaping how knowledge is created)

Organizations often occupy multiple spaces at once.

The key question is:

Where is your center of gravity moving?

Most organizations mistake movement along the vertical axis for transformation—adopting more AI tools—when the real shift requires movement across both axes, toward systemic integration and epistemic change.

Why Most Organizations Get Stuck

Most organizations plateau between Level 1 and Level 2.

They:

build AI capabilities

expand access

train users

But they do not redesign workflows or institutions.

Why?

Because moving to Level 3 and Level 4 requires:

structural change

role redefinition

governance redesign

It is no longer a technology problem.

It is an architectural one.[6]

The Risk of Medium-Level Power

As AI becomes a medium, its power increases—and so do its risks.

errors can propagate at scale

biases can become embedded in systems

shared models can create homogenized thinking

This introduces a new challenge:

epistemic risk at scale

Managing this requires:

traceability

continuous validation

contestability

diversity of models and perspectives

Governance becomes:

the design of how the system knows.

The Strategic Shift

This framework leads to a simple but critical shift:

From:

"Do we have AI?"

To:

"What kind of cognition are we building?"

Organizations that remain at Level 1 or 2:

improve efficiency

but preserve existing structures

Organizations that move to Level 3 and 4:

redesign how thinking happens

gain structural advantage

The difference is not adoption.
It is cognition.

Closing

AI maturity is not a story about technology.

It is a story about the evolution of cognition in systems.

The transition from tool to medium is the defining shift.

And it reframes the central question for leaders:

Not how to deploy AI,
but how to design the system in which intelligence
emerges.

That design will determine:

how decisions are made

how knowledge evolves

how institutions function

In the next chapter, we turn to one of the most visible
domains where this transformation is already unfolding:

Education.

Because when intelligence moves,
learning must move with it.

Endnotes

1. The relocation of cognition is already the book's central premise,
   so the more specific question here is what changes when AI
   becomes the environment of reasoning rather than an occasional
   instrument. Hollan, Hutchins, and Kirsh (2000) is the best
   conceptual bridge, and Amershi et al. (2019) is the best design
   bridge.

2. The distinction between tool and medium is clearer in media theory than in AI commentary. Marshall McLuhan's Understanding Media (1964) remains important because media reshape the conditions of thought rather than merely improving a task, and Merlin Donald's Origins of the Modern Mind (1991) shows the same point historically for representational systems.

3. On representational systems reshaping cognition, Donald (1991), Hutchins (1995), Zhang and Norman (1994), and Zhang (1997) belong together. The shared claim is that once representation changes, the space of possible reasoning changes with it. That is the conceptual core of this chapter.

4. The internet made information broadly available; generative AI makes representational work broadly available. Brown et al.'s 'Language Models are Few-Shot Learners' (2020) and OpenAI's GPT-4 Technical Report (2023) mark that shift at the model level, while Boiko et al.'s 'Autonomous Chemical Research with Large Language Models' (2023) shows it operating in a workflow.

5. When this chapter says that the system begins to think, it is speaking in the distributed-cognition sense already established earlier. Hutchins (1995) and Hollan, Hutchins, and Kirsh (2000) provide the conceptual frame, while Dell'Acqua et al.'s 'Navigating the Jagged Technological Frontier' (2026) shows how integrated human-AI workflows behave in practice.

6. The architecture-versus-adoption point is by now a recurring theme. David's 'The Dynamo and the Computer' (1990), Orlikowski's 'Using Technology and Constituting Structures' (2000), and Brynjolfsson, Rock, and Syverson's 'The Productivity J-Curve' (2021) all show that system redesign, not mere uptake, is what produces durable advantage.

CHAPTER 7

Education After Knowledge

For centuries, education has been organized around a simple premise:

Knowledge must be transferred into the minds of individuals.

Curricula are designed to deliver it.
Assessments are designed to measure it.
Credentials are awarded to certify it.

This model made sense in a world where knowledge was scarce, access was limited, and cognitive work depended on what one could recall and apply independently.

That world no longer exists.

When knowledge is external, dynamic, and instantly accessible, education cannot be organized around its transmission.

Yet most educational systems continue to operate as if nothing has changed.

They are not failing.
They are succeeding at the wrong task.

The Knowledge Transfer Model

Traditional education rests on three assumptions:

Knowledge is scarce and valuable
It must be internalized to be useful

Learning is demonstrated through independent performance

These assumptions shaped everything:

lectures to transmit information
textbooks to standardize content
exams to test recall and application

The goal was clear:

Fill the mind with knowledge, then test its retention and use.

This model produced generations of capable professionals. It aligned with the demands of an industrial and early digital economy.

But it is misaligned with a world where knowledge is no longer the limiting factor.

Education is not ending knowledge—it is moving beyond it as the limiting factor.

The Break: Knowledge Becomes External

AI systems fundamentally alter the role of knowledge in learning.

Knowledge is:

instantly accessible
continuously updated
dynamically generated

A student no longer needs to memorize large bodies of content to access them. A professional no longer relies solely on internal knowledge to perform complex tasks.

This does not mean knowledge is irrelevant.

It means its location has changed.

Knowledge has moved from the mind to the system.

And when the location changes, the function of education must change with it.

The Illusion of Academic Integrity

One of the first institutional reactions to AI has been to restrict its use in learning environments.

Policies attempt to:

> prohibit AI assistance
> detect AI-generated content
> preserve "independent work"

These responses are understandable. They aim to protect the integrity of assessment and the value of credentials.

But they are built on an assumption that is no longer valid:

> That learning can—or should—be separated from the tools through which cognition now occurs.

Prohibiting AI in education is like prohibiting writing to preserve memory.

It attempts to preserve a model of cognition that has already been transformed.

> The issue is not whether students use AI.
> It is whether we are designing learning for a world in which they inevitably will.

From Knowledge Acquisition to Cognitive Orchestration

If knowledge is external and cognition is distributed, then the purpose of education shifts.

From:

acquiring knowledge

To:

orchestrating cognition

This includes the ability to:

frame meaningful questions
engage productively with AI systems
evaluate outputs critically
integrate information across sources
make decisions under uncertainty

Learning becomes less about what you know and more about how you think within a system.

Education must move from teaching content to designing cognitive capability.

The Diagnostic Map for Education

Education systems can be understood using the same diagnostic introduced in the previous chapter.

Degree of integration (localized → systemic) — how deeply AI is embedded in learning workflows and environments.

Nature of impact (operational → epistemic) — whether AI changes task execution (e.g., faster writing) or fundamentally reshapes how knowledge is produced, evaluated, and applied.

Most institutions cluster in the upper-left quadrant: localized, operational use—tools added to existing assignments.

Progress requires movement along two axes:

From localized → systemic (integrating AI into curricula, assessment, and feedback loops)

From operational → epistemic (shifting from task completion to how knowledge is generated and judged)

The key question is not whether students use AI.

It is whether the system is designed for how they think with it.

Most schools are moving faster along the vertical axis—adopting tools—while avoiding the horizontal shift that would redefine learning itself.

The Transformation of Learning Tasks

As in other domains, the nature of cognitive tasks in education changes:

Memorization → Recognition
Problem solving → Problem framing
Writing → Iterative co-creation
Research → Evaluation and synthesis

These shifts do not reduce intellectual rigor.

They relocate it.

The New Core Skills

In an AI-mediated world, foundational skills are redefined.

Students must develop:

Cognitive Framing

Interaction Design

Critical Evaluation

Integration

Judgment

The core skill is no longer knowing.

It is knowing how to think within a system that thinks with you.

Rethinking Assessment

Assessment is where the misalignment becomes most visible.

Traditional exams test:

recall
independent reasoning
closed-book problem solving

But these conditions do not reflect real-world cognition, where:

information is accessible
tools are available
collaboration is the norm

We are measuring what is easy to test, not what matters.

Toward Authentic Cognitive Assessment

Assessment must evolve to reflect distributed cognition.

Evaluate:

how students work with AI systems
how they structure problems
how they evaluate and refine outputs
how they make decisions in complex scenarios

Examples:

open-system problem solving
iterative projects with AI collaboration
real-world simulations
evaluation of generated outputs

The Role of the Educator

From:

content provider
knowledge authority

To:

cognitive guide
system designer
evaluator of judgment

Educators design environments where thinking is orchestrated, not merely delivered.

The Institutional Challenge

Structures built for content delivery—courses, departments, credentials—become unstable when knowledge is external and dynamic.

The architecture of education must evolve from content delivery to cognitive development.

Beyond the Classroom

Learning becomes:

continuous rather than episodic
embedded in work rather than separate from it
driven by problems rather than curricula

A New Definition of Learning

Learning is the process of developing the ability to think effectively within distributed cognitive systems.

Transition

If education must be redesigned for distributed cognition, the same is true for another foundational domain:

Research.

When the bottleneck shifts from generating ideas to evaluating them, the nature of discovery changes.

The next chapter examines how research evolves when cognition is distributed—and what this means for the future of science.

Endnotes

1. The educational shift from knowledge transfer to cognitive orchestration is already latent in Lev Vygotsky's Mind in Society (1978), Jean Lave's Cognition in Practice (1988), and Brown, Collins, and Duguid's 'Situated Cognition and the Culture of Learning' (1989). Read with Hutchins's Cognition in the Wild (1995), they suggest that learning is better understood as guided participation in systems of thought than as simple content deposit.

2. On externalized knowledge, Evan Risko and Sam Gilbert's 'Cognitive Offloading' (2016) is the cleanest modern synthesis, and Zhang's 'The Nature of External Representations in Problem Solving' (1997) remains crucial for showing why offloading changes the structure of cognition rather than merely lightening memory load.

3. The key issue in AI and learning is less whether AI is present than how the interaction is designed. Saleema Amershi et al.'s 'Guidelines for Human-AI Interaction' (2019) is useful here because it treats explanation, timing, uncertainty, and iteration as design variables. That is a better guide for this chapter than blanket pro- or anti-AI classroom rhetoric.

4. On the limits of generative models, Bender et al.'s 'On the Dangers of Stochastic Parrots' (2021) and OpenAI's GPT-4 Technical Report (2023) make the point plainly: fluency is not the same as grounded understanding. In education, that means students need stronger habits of evaluation, not weaker standards.

5. The assessment problem has a long prehistory. Grant Wiggins's Educative Assessment (1998), Lave's Cognition in Practice (1988), and Brown, Collins, and Duguid (1989) all argue, in different ways, that tests detached from authentic contexts can measure the wrong thing. This chapter extends that argument into human-AI settings.

6. The real unit of educational performance is increasingly the human-AI system. McDuff et al.'s 'Towards Accurate Differential Diagnosis with Large Language Models' (2025) is not an education paper, but it is useful because it shows how a well-structured conversational system can improve expert reasoning in practice; Dell'Acqua et al. (2026) shows, just as importantly, that poor fit between task and model can degrade outcomes. The educational analogue is straightforward.

CHAPTER 8

Research After Discovery

For centuries, research has been organized around a single, central activity:

The generation of new knowledge.

Scientists were trained to:

observe carefully

formulate hypotheses

design experiments

interpret results

The process was sequential, deliberate, and constrained by human cognitive limits. Progress depended on the ability of individuals and small teams to generate insight from limited data and limited tools.

This model produced extraordinary advances.

But it was built around a fundamental constraint:

The scarcity of ideas.

That constraint is now dissolving.

This does not mean discovery ends.
It means discovery is no longer the constraint.

The End of Idea Scarcity

Artificial intelligence fundamentally changes the economics of discovery.

AI systems can now:

generate hypotheses

propose experimental designs

identify patterns in large datasets

simulate outcomes

synthesize literature across domains

What once required years of training and effort can now be initiated in seconds.

This does not mean AI replaces researchers.

It means that one of the core bottlenecks of research—idea generation—has been dramatically reduced.

When ideas become abundant, they cease to be the limiting factor.[1]

Discovery does not disappear.
It becomes abundant—and therefore no longer defines research.

A recent wave of "AI co-scientist" systems makes this shift concrete: models that read across vast literatures, propose experimentally testable hypotheses, and even suggest candidate therapeutics. In some cases, AI-discovered leads have advanced to clinical testing. The deeper point is not any single result—it is that cognition is relocating within the research process.

For decades, the structure of science was clear: humans generated ideas and machines analyzed data. That boundary is dissolving. We are moving toward machine-generated discovery with human oversight—systems that continuously generate, evaluate, and refine hypotheses within shared workflows.

The New Bottleneck: Evaluation

As the cost of generating ideas approaches zero, the constraint shifts.

From:

producing hypotheses

To:

evaluating them

This shift is structural.

A single researcher, working with AI, can now generate:

dozens of plausible hypotheses

multiple experimental pathways

alternative interpretations

But not all hypotheses are:

valid

meaningful

testable

worth pursuing

The challenge is no longer to generate ideas.

It is to decide which ideas matter.

Evaluation becomes the central function of research.[2]

This creates an emerging ideation–execution gap: when AI can generate thousands of plausible hypotheses, the constraint shifts to what we test, how quickly we can validate, and who decides what matters.

From Pipeline to Loop

Traditional research follows a pipeline:

Formulate hypothesis → Design experiment → Collect data → Analyze results → Publish findings

This model assumes:

hypotheses are scarce

experiments are costly

iteration is slow

In an AI-mediated environment, this pipeline becomes a loop:

hypotheses are generated continuously

simulations precede physical experiments

data is analyzed in real time

feedback loops refine direction

Research becomes:

parallel rather than sequential

exploratory rather than linear

system-driven rather than individual-driven

Discovery is no longer a pipeline.[3]

It is a continuous system.

The Role of Simulation

Simulation becomes a central mechanism of discovery.

AI enables:

virtual experimentation

model-based prediction

large-scale scenario exploration

This does not replace empirical science.

It restructures it.

Instead of exploring vast hypothesis spaces physically, researchers:

narrow the space computationally

test high-value hypotheses empirically

Simulation compresses the search space of discovery.[4]

The next phase is already visible: AI systems connected to automated labs, enabling continuous loops of hypothesis → experiment → refinement, running at a cadence far beyond traditional cycles.

The Expansion of the Hypothesis Space

AI dramatically expands the space of possible ideas.

It can:

combine concepts across domains

surface non-obvious relationships

explore configurations beyond human intuition

This creates opportunity—but also noise.

Many generated hypotheses will be:

trivial

incorrect

uninterpretable

More ideas do not produce better science.

They require better selection.[5]

The Changing Role of the Researcher

As structure changes, so does the role of the researcher.

From:

generator of hypotheses

designer of experiments

analyzer of data

To:

curator of hypothesis spaces

designer of evaluation systems

interpreter of system outputs

The researcher becomes an orchestrator.

Creativity shifts from producing ideas to shaping the space in which ideas are generated and evaluated.

Distributed Discovery

Research becomes distributed across:

human researchers

AI systems

data infrastructures

collaborative networks

No single element controls discovery.

Discovery emerges from interaction.

Science becomes a distributed cognitive system.[6]

Rethinking Originality

In the traditional model, originality is tied to:

novelty of ideas

uniqueness of hypotheses

In an AI-mediated system, originality shifts to:

identifying meaningful questions

framing problems effectively

selecting high-impact directions

integrating across domains

Originality is no longer about having ideas others do not have.

It is about recognizing which ideas matter.

The Risk of Superficial Discovery

As generation becomes easy, superficial output increases.

Risks include:

overproduction of low-value studies

amplification of spurious correlations

erosion of signal within the literature

The challenge is not technical.

It is epistemic.

When generation scales, validation must scale.[7]

The Need for Evaluation Systems

Research must develop stronger evaluation architectures.

These include:

  robust validation methods

  reproducibility standards

  transparent AI-assisted workflows

  integration of human judgment with machine outputs

  Evaluation must become:

  systematic

  scalable

  aligned with generative capacity[8]

Collapsing Disciplinary Boundaries

AI enables cross-domain integration by:

  translating representations

  linking datasets

  identifying shared structures

  Disciplines become:

  more porous

  less bounded

  Problems become more interconnected.

  Discovery becomes integrative.

Acceleration and Compounding

Research accelerates through feedback loops.

faster iteration

continuous refinement

compounding capability

Better tools produce faster discovery.

Faster discovery produces better tools.

Science becomes a continuously learning system.

Institutional Implications

Research institutions remain organized around the old model:

discrete grants

publication outputs

individual investigators

These structures misalign with:

continuous discovery

distributed cognition

AI-mediated workflows

Institutions must rethink:

funding models

collaboration structures

evaluation metrics

incentive systems[9]

A New Definition of Research

Research can now be redefined:

Research is the process of navigating, evaluating, and refining an expanded space of possible knowledge within a distributed cognitive system.

It integrates:

human judgment

machine generation

iterative validation

It is not less rigorous.

It is more complex—and more powerful.

Transition

If research is no longer defined by generating ideas, and expertise is no longer defined by possessing knowledge, then another domain must be reconsidered:

Decision-making.

When cognition is distributed and reasoning is externalized, decisions can no longer rely on individual judgment alone.

The next chapter examines how decisions are made in human–AI systems—and why leadership becomes the design of cognition rather than the exercise of authority.

Endnotes

1. The chapter's claim that discovery is moving from idea scarcity to idea abundance is most directly visible in Chris Lu et al.'s 'The AI Scientist: Towards Fully Automated Open-Ended Scientific Discovery' (2024) and Juraj Gottweis et al.'s 'Towards an AI Co-Scientist' (2025). Whatever one thinks of the systems' maturity, they make the structural point unmistakable: hypothesis generation is no longer the rarest part of the pipeline.

2. The ideation-execution gap appears as soon as generation outruns validation. Andres M. Bran et al.'s 'Augmenting Large Language Models with Chemistry Tools' (2024) shows how many plausible paths can be opened quickly, while Nathan J. Szymanski et al.'s 'An Autonomous Laboratory for the Accelerated Synthesis of Inorganic Materials' (2023) reminds us that physical testing remains the harder bottleneck.

3. For the shift from pipeline to loop, the clearest examples are the A-Lab paper by Szymanski et al. (2023) and Boiko et al.'s 'Autonomous Chemical Research with Large Language Models' (2023). Both replace the old serial picture of research with iterative cycles of proposal, experiment, interpretation, and refinement.

4. The AlphaFold result is still the canonical example of simulation compressing the search space. John Jumper et al.'s 'Highly Accurate Protein Structure Prediction with AlphaFold' (2021) matters not just because it is accurate, but because it changes which laboratory work is worth doing next.

5. On expanded hypothesis spaces, Renqian Luo et al.'s 'BioGPT: Generative Pre-trained Transformer for Biomedical Text Generation and Mining' (2022) is a good marker because it shows what happens when a generative model is trained deeply inside a scientific domain. Gottweis et al.'s 'Towards an AI Co-Scientist' (2025) pushes the same logic toward hypothesis formation and ranking.

6. The phrase 'AI co-scientist' is now literal enough to cite directly. Boiko et al.'s 'Autonomous Chemical Research with Large Language Models' (2023) and Gottweis et al.'s 'Towards an AI Co-Scientist' (2025) show systems that do more than summarize papers or retrieve facts; they generate proposals, use tools, and help steer experimental agendas.

7. The risk of superficial discovery grows whenever output scales faster than scrutiny. The AI Scientist paper (2024) is useful here precisely because it makes the possibility vivid, while Brian Nosek et al.'s 'Promoting an Open Research Culture' (2015) and Marcus Munafo et al.'s 'A Manifesto for Reproducible Science' (2017) explain why scientific value depends on verification, transparency, and selection pressure, not on sheer volume.

8. For evaluation architectures, the strongest anchors are Nosek et al. (2015), Munafo et al. (2017), and Mark Wilkinson et al.'s 'The FAIR Guiding Principles for Scientific Data Management and Stewardship' (2016). Together they capture the institutional side of this chapter's argument: if generation accelerates, standards for validation, traceability, and reuse have to strengthen with it.

9. On institutional misalignment, the most authoritative overview is the OECD volume Artificial Intelligence in Science (2023). It is especially useful because it connects technical possibilities to funding, infrastructure, skills, and governance - exactly the layer where this chapter says the next bottlenecks will appear.

CHAPTER 9

Decision-Making After Expertise

For most of human history, decision-making has been anchored in a simple premise:

Those who know more decide.

Authority followed knowledge. Expertise justified power. Leaders were expected to possess superior understanding, accumulated through experience, training, and judgment.

This model worked when knowledge was scarce and cognition was bounded by the individual.

It does not hold in a world where cognition is distributed.

When intelligence is no longer located within individuals, decision-making cannot be either.

Expertise does not disappear.
It becomes distributed—and therefore no longer defines decision-making.

Artificial intelligence does not just change how decisions are made.
It changes where decisions are made—and who, or what, participates in them.[1]

The Old Model: Decisions Inside the Head

Traditional decision-making assumes:

information is gathered

analyzed internally

weighed against experience

translated into action

Even when groups are involved, the process ultimately resolves within human minds. Tools may support analysis, but they do not fundamentally alter the locus of cognition.

The decision-maker remains the center.

This model shaped:

executive leadership

clinical judgment

policy-making

operational management

The core expectation was clear:

Good decisions come from better thinkers.

The Break: Reasoning Becomes External

AI systems fundamentally alter this process.

They can:

generate options

simulate scenarios

predict outcomes

identify patterns across vast datasets

Much of the reasoning that once occurred internally can now occur externally.

A leader does not begin with a blank mental model.
A clinician does not rely solely on internal knowledge.

A policymaker does not manually synthesize all available data.

Instead, they engage with systems that:

propose alternatives

highlight trade-offs

surface risks

Decision-making becomes a process of interacting with external reasoning.[2]

From Decision-Maker to Decision System

As cognition becomes distributed, the unit of decision-making shifts.

From:

individual decision-maker

To:

human–AI decision system

In this system:

humans provide context, values, and judgment

AI provides generation, analysis, and prediction

Neither alone produces the final decision.

Decisions emerge from the interaction.[3]

This has a critical implication:

The quality of decisions depends not only on the intelligence of the human or the capability of the AI, but on how well the system is structured.

The Explosion of Options

One immediate effect of AI is the expansion of the decision space.

Where once a decision-maker considered a limited set of options, AI can generate:

numerous alternatives

variations of strategies

different scenario projections

This creates both opportunity and challenge.

More options increase the likelihood of finding better solutions.
But they also increase complexity.

More choices do not simplify decisions.
They make them harder.

The bottleneck shifts again:

From:

generating options

To:

selecting among them.[4]

Judgment Under Abundance

In an environment of abundant possibilities, judgment becomes the defining capability.[5]

Judgment includes:

prioritizing options

understanding trade-offs

aligning decisions with values and goals

acting under uncertainty

AI can inform judgment.
It cannot fully replace it.

Decision-making becomes less about finding answers and more about choosing among them.

The Illusion of Objectivity

AI introduces a new perception: that decisions can become more objective.

Models produce:

data-driven insights

probabilistic predictions

structured analyses

This can create the impression that decisions are:

neutral

unbiased

purely rational

But this is an illusion.[6]

AI systems are shaped by:

training data

model assumptions

design choices

Human judgment remains essential to:

interpret outputs

contextualize recommendations

identify limitations

AI does not remove subjectivity.
It relocates and transforms it.

Distributed Responsibility

As decisions become distributed across human–AI systems, responsibility becomes more complex.

When a decision leads to failure:

is it the human's fault?

the model's fault?

the system's design?

Traditional accountability frameworks assume:

clear decision-makers

traceable reasoning

human control

But in distributed systems:

reasoning is shared

contributions are intertwined

causality is less transparent

Responsibility becomes a system property.[7]

This creates new challenges for:

governance

ethics

regulation

Speed vs. Reflection

AI accelerates decision-making.

analysis is faster

options are generated instantly

feedback loops are shortened

This creates pressure to:

    decide more quickly

    act more frequently

    But speed introduces risk.

    rapid decisions may bypass reflection

    over-reliance on AI outputs may reduce scrutiny

    complexity may outpace understanding

    Faster decisions are not necessarily better decisions.[8]

    The challenge is to balance:

speed

depth

judgment

The New Role of Leadership

As decision-making changes, so does leadership.

Traditional leadership emphasized:

    knowledge

    experience

    authority

In distributed cognitive systems, leadership shifts toward:

Framing
Defining the problem space and objectives

System design
Structuring how humans and AI interact

Judgment
Making final decisions under uncertainty

Alignment
Ensuring decisions reflect values and goals

Leaders become architects of decision systems.[9]

They are no longer the sole source of answers.
They are responsible for how answers are generated, evaluated, and acted upon.

The Risk of Over-Reliance

As AI becomes more capable, there is a risk of over-reliance.

Decision-makers may:

defer too quickly to model outputs

accept plausible answers without sufficient scrutiny

lose confidence in their own judgment

This creates a paradox:

AI increases capability

but can reduce engagement

The danger is not that AI will replace decision-makers.
It is that decision-makers will disengage.

Maintaining active judgment becomes critical.[10]

Designing Decision Systems

To function effectively, human–AI decision systems must be intentionally designed.

This includes:

Transparency
Understanding how outputs are generated

Feedback loops
Enabling continuous refinement

Role clarity
Defining what humans and AI are responsible for

Checks and balances
Preventing over-reliance or misuse

Decision-making becomes less about individual skill and more about system design.[11]

A New Definition of Decision-Making

We can redefine decision-making:

Decision-making is the process of coordinating human judgment and machine reasoning within a distributed cognitive system to select and act under uncertainty.

This includes:

generation

evaluation

selection

execution

It is not simpler than traditional decision-making.

It is more complex—and more powerful.

Transition

If decision-making is no longer centered on individuals, then governance—the system that structures authority, accountability, and coordination—must also evolve.

Most governance systems were designed for:

human decision-makers

clear hierarchies

bounded cognition

They are not designed for:

distributed cognition

human–AI systems

continuous, adaptive decision processes

The next chapter examines how governance must be reimagined in the cognitive era—and why it may be the most critical bottleneck in the transition to AI-native institutions.[1][2]

Endnotes

1. The distributed-decision frame has already been built earlier, so the more specific bridge here is from that literature into interactive AI assistance. Hutchins's Cognition in the Wild (1995) remains the conceptual base, and Daniel McDuff et al.'s 'Towards Accurate Differential Diagnosis with Large Language Models' (2025) shows how reasoning quality changes when a professional decision-maker works with an interactive model rather than in isolation.

2. On externalized reasoning, OpenAI's GPT-4 Technical Report (2023) marks the capability shift, but McDuff et al. (2025) makes the decision implication clearer: the model can expand and improve a clinician's differential, yet the clinician still has to interpret, filter, and own the result.

3. The decision system idea is best anchored in Saleema Amershi et al.'s 'Guidelines for Human-AI Interaction' (2019) and Fabrizio Dell'Acqua et al.'s 'Navigating the Jagged Technological Frontier' (2026). Both suggest that the relevant performance variable is not the human or the model taken alone, but the quality of coordination between them.

4. Decision-space expansion is one of the practical consequences of generative models. Brown et al.'s 'Language Models are Few-Shot Learners' (2020) and GPT-4 (2023) show why models can produce many plausible continuations, options, and reframings at low cost. The leadership burden then shifts from inventing candidates to selecting among them.

5. For judgment under uncertainty, Amos Tversky and Daniel Kahneman's 'Judgment under Uncertainty: Heuristics and Biases' (1974) is still foundational. It is helpful here because more options do not remove bias or ambiguity; they often increase the need for disciplined judgment.

6. The illusion of objectivity is well treated in Solon Barocas and Andrew Selbst's 'Big Data's Disparate Impact' (2016) and Andrew Selbst et al.'s 'Fairness and Abstraction in Sociotechnical Systems' (2019). Both show that apparently neutral outputs inherit assumptions from data, framing, and institutional context.

7. Responsibility becomes harder to assign when outcomes are co-produced by people, models, and workflows. Joshua Kroll et al.'s 'Accountable Algorithms' (2017) and Selbst et al. (2019) are useful here because they move accountability away from a simple model-only story and back toward the larger system.

8. On speed versus reflection, the older human-automation literature still holds up. Raja Parasuraman and Victor Riley's 'Humans and Automation: Use, Misuse, Disuse, Abuse' (1997) and John Lee and Katrina See's 'Trust in Automation' (2004) both explain why faster automated support can worsen decisions when trust outruns understanding.

9. The chapter's claim that leaders become designers of decision systems is closer to Herbert Simon's The Sciences of the Artificial (1969) and Karl Weick's Sensemaking in Organizations (1995) than to conventional leadership literature. Simon frames

organizations as designed systems; Weick shows how meaning and action emerge through ongoing interpretive processes.

10. Over-reliance is not a new AI problem so much as an intensified automation problem. Parasuraman and Riley (1997) introduced the classic vocabulary of misuse, disuse, and abuse, and Lee and See (2004) showed why calibrated trust is central. This chapter is extending that warning into model-mediated decision-making.

11. For decision-system design, Amershi et al. (2019) and Lee and See (2004) belong together. One gives practical guidelines for human-AI interaction; the other explains the trust conditions under which such guidelines matter.

12. The institutional consequences point forward into governance. NIST's AI Risk Management Framework (2023) and the OECD AI Principles (2019, revised 2024) are useful because they both assume that AI performance and AI risk have to be managed at the system level rather than left to individual discretion.

CHAPTER 10

Governance in the Cognitive Era

If cognition has moved beyond the individual, and decision-making now occurs within human–AI systems, then governance—the structure that defines authority, accountability, and control—must change as well.[1]

Yet most governance systems have not.

They remain anchored in assumptions that no longer hold:

> that decisions are made by identifiable individuals
>
> that reasoning is contained and traceable
>
> that authority flows through stable hierarchies

These assumptions were appropriate for a world in which cognition was bounded and slow.

They are misaligned with a world in which cognition is distributed, dynamic, and partially external.

> Control does not disappear.
> It becomes distributed—and therefore must be redesigned.

Governance has become the bottleneck of the cognitive era.

> Governance is no longer the control of decisions.
> It is the design of cognition.[2]

The Old Model: Control Through Hierarchy

Traditional governance is built on three principles:

Authority is centralized

Responsibility is assigned to individuals or roles

Processes are designed for predictability and control

This model assumes:

decisions can be reviewed

actions can be traced

outcomes can be attributed

It is designed to reduce uncertainty by:

slowing processes

standardizing decisions

enforcing compliance

In stable environments, this works.

In rapidly evolving, cognitively distributed environments, it creates friction.[3]

The Breakdown of Traditional Assumptions

Distributed cognition challenges each of these principles.

Authority Becomes Diffuse

When decisions emerge from human–AI systems, authority is no longer fully centralized.

The system contributes.
The human interprets.
The outcome is co-produced.

Responsibility Becomes Ambiguous

Who is responsible when:

an AI-generated recommendation is followed

an error arises from system interaction

a decision reflects both human and machine input

Traditional accountability frameworks struggle to assign responsibility in these contexts.[4]

Processes Become Too Slow

Governance structures designed for deliberation and control cannot keep pace with:

rapid iteration

continuous data flows

real-time decision-making

Governance designed for stability struggles in environments defined by change.

From Control to Enablement

In the cognitive era, governance must shift its primary function.

From:

controlling decisions

To:

enabling effective decision systems

This includes:

creating conditions for high-quality human–AI interaction

defining boundaries and guardrails

ensuring alignment with values and objectives

Governance is no longer about restricting action.
It is about shaping how cognition occurs.[5]

The New Governance Questions

Traditional governance asks:

Who decides?

What are the rules?

How do we ensure compliance?

Cognitive-era governance must ask:

How are decisions generated within the system?

How do humans and AI interact?

How do we ensure alignment and reliability?

How do we monitor and adapt continuously?

These are not static questions.
They require ongoing adjustment.

Designing Guardrails, Not Gatekeepers

In hierarchical systems, governance often operates through
gatekeeping:

approvals

reviews

checkpoints

This approach assumes:

decisions are discrete

processes are linear

In distributed cognitive systems:

decisions are continuous

processes are iterative

Gatekeeping becomes impractical.

Instead, governance must rely on guardrails:

constraints that guide behavior

principles that shape interaction

boundaries that define acceptable outcomes

Guardrails enable speed while maintaining alignment.[6]

The Role of Transparency

As AI participates in cognition, transparency becomes critical.

Not full technical transparency—which may be infeasible—but functional transparency:

understanding what the system is doing

recognizing its limitations

interpreting its outputs

Without transparency:

trust erodes

misuse increases

errors propagate

Governance must ensure that participants in the system can:

understand

question

challenge

the outputs of AI.[7]

Continuous Oversight, Not Periodic Review

Traditional governance relies on periodic review:

audits

evaluations

compliance checks

In fast-moving systems, this is insufficient.

Governance must become:

continuous

embedded

adaptive

This includes:

real-time monitoring

feedback loops

iterative policy adjustment

Oversight must operate at the same speed as cognition.[8]

The Emergence of AI Governance Structures

New governance structures are beginning to appear:

AI councils

ethics committees

oversight boards

But many of these risk becoming part of innovation theater if they:

lack authority

operate in isolation

focus on policy without integration

Effective governance structures must be:

integrated into core operations

connected to decision-making processes

empowered to act

Balancing Innovation and Risk

Governance must navigate a fundamental tension:

enabling innovation

managing risk

Too much control:

slows progress

discourages adoption

Too little control:

increases errors

undermines trust

The goal is not to eliminate risk, but to manage it intelligently.[9]

Distributed Governance

As cognition becomes distributed, governance must also become distributed.

This means:

decisions are guided at multiple levels

responsibility is shared

oversight is embedded across the system

Central governance still plays a role, but it cannot manage all decisions directly.

Governance becomes a system property, not a single function.[10]

The Human Role in Governance

Despite increasing automation, humans remain essential to governance.

They provide:

values

ethical judgment

contextual understanding

AI can assist with:

monitoring

analysis

detection of anomalies

But governance ultimately depends on human judgment.

A New Definition of Governance

We can redefine governance:

Governance is the design and continuous adaptation of structures, processes, and constraints that enable distributed cognitive systems to operate effectively, responsibly, and in alignment with organizational values.[11]

This includes:

enabling action

guiding interaction

ensuring accountability

adapting to change

The Institutional Imperative

Organizations that fail to evolve governance will face:

misalignment between capability and control

increased risk of errors and misuse

inability to scale AI effectively

Conversely, those that redesign governance will:

move faster

make better decisions

maintain trust

Governance is not a constraint on transformation. It is a prerequisite for it.

Transition

We have now examined how AI reshapes:

cognition

expertise

education

research

decision-making

governance

Each domain reveals the same pattern:

Cognition is no longer located within individuals.
It is distributed across systems.

This leads to a final, unifying question:

If all core functions are being reorganized around distributed cognition, what does it mean to design an institution from the ground up for this reality?

The next chapter brings these threads together—moving from analysis to design.

From understanding the cognitive revolution to building for it.

This is the blueprint for AI-native institutions.

Governance is no longer the control of decisions.
It is the design of cognition.

Endnotes

1. The governance challenge begins where the distributed-decision problem leaves off. McDuff et al. (2025) and Dell'Acqua et al. (2026) are helpful here because they show, in different ways, that outcomes in AI-assisted work are joint products. Once that is true, governance can no longer assume a single, self-contained decision-maker.

2. The most authoritative recent framing of governance as a practical bottleneck comes from NIST's AI Risk Management Framework (2023) and the OECD AI Principles (2019, revised 2024). Both are notable for how explicitly they move from model quality to organizational process, oversight, and context.

3. On hierarchical control systems, James March and Herbert Simon's Organizations (1958) still matters because it treats formal organizations as bounded systems of authority, routines, and information processing. This chapter's point is that those assumptions strain once cognition becomes distributed and partially externalized.

4. For distributed responsibility, Joshua Kroll et al.'s 'Accountable Algorithms' (2017) and Andrew Selbst et al.'s 'Fairness and Abstraction in Sociotechnical Systems' (2019) are the strongest anchors. They both show why accountability fails when we isolate the technical artifact from the institutional setting in which it operates.

5. Governance as system design is already implicit in Wanda Orlikowski's 'Using Technology and Constituting Structures' (2000), and it becomes explicit in NIST's AI Risk Management Framework (2023). The underlying idea is that governance works best when it shapes practice continuously, not when it intervenes only after the fact.

6. The guardrails-versus-gatekeepers distinction is very close to the spirit of NIST AI RMF (2023) and the OECD AI Principles (2019, revised 2024). These frameworks do not assume that safe AI comes from freezing action; they assume it comes from defining boundaries, monitoring risk, and keeping humans able to intervene.

7. On transparency, Joshua Kroll et al. (2017) is useful because it asks what kinds of account-giving are possible for algorithmic systems, while NIST AI RMF (2023) treats explainability and traceability as operational governance properties rather than abstract ideals.

8. The case for continuous oversight is increasingly explicit in current public-sector AI guidance. NIST AI RMF (2023) provides the risk-management vocabulary, and the OECD report Governing with Artificial Intelligence (2025) shows how pilot-heavy, episodic governance often fails to scale because measurement, authority, and feedback are too weak.

9. The claim that risk must be managed rather than eliminated is built directly into NIST AI RMF (2023). That is one reason the framework is more useful for this chapter than generic ethics

language: it assumes real organizations must keep acting while learning to manage uncertainty.

10. On distributed governance, Elinor Ostrom's Governing the Commons (1990) is still instructive because it shows how complex systems can be governed through layered, distributed arrangements rather than single-point command. The analogy is not exact, but the organizational lesson travels well.

11. Governance redesign is where the policy and organizational literatures converge. NIST AI RMF (2023), the OECD AI Principles (2019, revised 2024), and OECD's Governing with Artificial Intelligence (2025) all point toward the same conclusion: AI cannot be scaled responsibly with legacy governance structures left untouched.

CHAPTER 11

The Architecture of AI-Native Institutions

We have examined the transformation of cognition across domains:

> thinking has moved from the individual to the system

> expertise has shifted from possession to judgment

> education has moved from knowledge transfer to cognitive orchestration

> research has shifted from generation to evaluation

> decision-making has become distributed

> governance has become the central bottleneck

> Each change points to the same conclusion:

> The underlying architecture of our institutions is misaligned with how cognition now works.[1]

> Most organizations were designed for a world in which:

> intelligence resided in individuals

> knowledge was scarce

> decisions were localized

> change was slow

> That world has ended.

What replaces it is not simply a more advanced version of the same institutions.

It is a different kind of institution altogether.

An AI-native institution is not one that uses AI.
It is one that is designed around distributed cognition.[2]

From Organizations to Cognitive Systems

Traditional organizations are structured around:

roles

functions

hierarchies

workflows

These elements coordinate human activity.

In an AI-native institution, the primary design problem is different.

It is not how to organize people.
It is how to organize cognition.[3]

Institutions become cognitive systems.

This means designing how:

information flows

decisions are generated

humans and AI interact

feedback loops operate

The organization is no longer just a social structure.
It is a system for thinking at scale.

The Four Pillars of Cognitive Architecture

Across domains, institutional functions can be reduced to four core cognitive activities:

Learning — acquiring and updating knowledge

Discovery — generating new knowledge

Decision-making — selecting and acting

Governance — guiding and constraining

In traditional institutions, these functions are:

separated

slow-moving

human-centered

In AI-native institutions, they are:

integrated

continuous

distributed across human–AI systems[4]

The architecture must be redesigned across all four pillars.

Pillar 1 — Learning as Continuous Cognition

In AI-native institutions, learning is no longer confined to classrooms or training programs.

It becomes:

continuous

embedded in workflows

driven by real-time interaction with AI

Employees, students, and practitioners:

learn while working

receive adaptive feedback

access knowledge on demand

The boundary between learning and doing dissolves.

Learning becomes a property of the system, not a separate activity.[5]

Pillar 2 — Discovery as a System Process

Discovery is no longer limited to research departments.

AI enables:

hypothesis generation across the organization

data-driven insights in real time

continuous experimentation

Innovation becomes:

distributed

iterative

embedded in operations

Every part of the organization becomes a site of discovery.[6]

Pillar 3 — Decision-Making as a Distributed System

Decisions are no longer made solely by individuals or committees.

They emerge from:

human judgment

AI-generated insights

data-driven feedback

Effective decision systems require:

clear roles for humans and AI

structured interaction

mechanisms for evaluation and refinement

Decision-making becomes a designed system, not an individual act.[7]

Pillar 4 — Governance as Adaptive Control

Governance shifts from:

static rules

hierarchical control

To:

dynamic guardrails

continuous oversight

adaptive policies

It must:

operate at the speed of cognition

manage distributed responsibility

maintain alignment with values

Governance becomes an active, evolving system.[8]

Integration: The Continuous Loop

In AI-native institutions, these four pillars are not separate.

They form a continuous loop:

learning informs discovery

discovery informs decisions

decisions generate new data

governance shapes all stages

AI accelerates and connects these loops.

The institution becomes a continuously learning system.[9]

Redesigning Workflows

To operationalize this architecture, workflows must be redesigned.

Traditional workflows:

linear

segmented

human-centered

AI-native workflows:

iterative

integrated

human–AI collaborative

The process is continuous, not discrete.[10]

Redefining Roles

Roles within the organization must also change.

From:

task execution

knowledge possession

To:

system orchestration

judgment

interaction design

Roles are defined by how they contribute to the cognitive system.[11]

Data as Cognitive Infrastructure

Data becomes central to the architecture.

Not as a static asset, but as:

a dynamic input to cognition

a feedback mechanism

a driver of learning and discovery

Data is not a resource.
It is part of the thinking process.[12]

Technology as Infrastructure

AI must be treated as infrastructure, not as a tool.

It is not an add-on.

It shapes how the institution operates at every level.[13]

Cultural Transformation

Structural change must be accompanied by cultural change.

Organizations must shift from:

valuing knowledge possession

to valuing judgment and learning

Culture must align with distributed intelligence.[14]

The Transition Challenge

Moving to an AI-native architecture is not a simple upgrade.

It is a structural transformation.[15]

The Competitive Advantage

The advantage will not come from having AI.

It will come from being designed for it.[16]

A New Definition of Institution

An institution is no longer an organization of people.
It is a system for organizing cognition.

An institution is a system for organizing cognition at scale.[17]

Transition

Leaders can no longer rely solely on authority or individual judgment.

They must become architects of cognition.

Endnotes

1. The institutional misalignment problem is already familiar from earlier chapters, but here it becomes architectural. David's 'The Dynamo and the Computer' (1990) and Brynjolfsson, Rock, and Syverson's 'The Productivity J-Curve' (2021) remain the clearest explanation of why new cognitive infrastructure produces little until institutions reorganize around it.

2. There is not yet a settled canonical literature on 'AI-native institutions' as such. The closest foundations are Herbert Simon's The Sciences of the Artificial (1969), Paul David (1990), and Brynjolfsson, Rock, and Syverson (2021). Taken together, they

imply that once cognition becomes infrastructural, institutional form has to change with it.

3. On organizations as cognitive systems, the closest direct sources are Hutchins's Cognition in the Wild (1995) and Hollan, Hutchins, and Kirsh's 'Distributed Cognition' (2000). What they provide is a unit of analysis larger than the individual and precise enough to talk about workflows, representations, and coordination.

4. The integration of learning, discovery, and decision-making has affinities with James March's 'Exploration and Exploitation in Organizational Learning' (1991). The old institutional tendency is to separate these functions; the new one is to connect them through faster feedback and shared cognitive infrastructure.

5. On continuous learning systems, Chris Argyris and Donald Schon's Organizational Learning (1978) and Peter Senge's The Fifth Discipline (1990) remain strong anchors. They matter here because AI-native institutions learn less through periodic training and more through continuous correction inside work.

6. The discovery pillar is no longer hypothetical. Chris Lu et al.'s 'The AI Scientist' (2024) and Juraj Gottweis et al.'s 'Towards an AI Co-Scientist' (2025) show the beginnings of research systems in which hypothesis generation is embedded across an ongoing computational workflow.

7. The distributed-decision pillar is likewise becoming concrete. Amershi et al. (2019), Daniel McDuff et al. (2025), and Dell'Acqua et al. (2026) all show that decision quality depends on the structure of the human-AI system, not on the isolated excellence of either component.

8. On governance evolution inside such institutions, NIST AI RMF (2023) and the OECD AI Principles (2019, revised 2024) are the right contemporary references because both assume continuous adaptation rather than one-time rule setting.

9. The loop logic is visible both in organizational learning and in autonomous experimentation. March (1991) gives the learning language; Nathan J. Szymanski et al.'s A-Lab paper (2023) gives a concrete scientific example of continuous propose-test-refine cycles.

10. Workflow redesign has been the hidden issue all along. Orlikowski (2000) shows why technology only matters through changed practice, and David (1990) shows why those changes are often delayed until complementarities are rebuilt.

11. Role transformation follows from that same redesign. Dell'Acqua et al. (2026) is useful here because it shows that AI changes the content of knowledge work unevenly, which is exactly why new roles emerge around orchestration, verification, and judgment rather than around routine production alone.

12. On data as cognitive infrastructure, Hutchins (1995) remains conceptually helpful, while Wilkinson et al.'s 'The FAIR Guiding Principles' (2016) captures the infrastructural requirement more explicitly. Data becomes part of cognition only when it can move, be interpreted, and be recombined across the system.

13. The AI-as-infrastructure claim rests on the same general-purpose-technology logic used earlier. Bresnahan and Trajtenberg (1995) and Brynjolfsson, Rock, and Syverson (2021) both show why deep value comes from complementary redesign rather than isolated deployment.

14. Cultural change is not an afterthought here. Argyris and Schon (1978) and Senge (1990) both remind us that institutions do not become learning systems simply because new tools arrive; norms of inquiry, correction, and shared understanding have to change as well.

15. On transformation challenges, the historical lesson is again that pilots can persist without architectural consequences. David (1990), Brynjolfsson et al. (2021), and OECD's Governing with Artificial Intelligence (2025) all speak to that point from different angles.

16. Competitive advantage in this setting is less about first access than about better complements. That is the enduring lesson of David (1990) and Brynjolfsson et al. (2021), and it is the strategic logic behind the term 'AI-native' in this chapter.

17. The final claim that institutions organize cognition at scale is best read as a synthesis of Hutchins's Cognition in the Wild (1995) and Simon's The Sciences of the Artificial (1969). One gives the distributed-cognition lens; the other gives the design lens.

CHAPTER 12

Leadership When No One Thinks Alone

For most of modern history, leadership has been defined by a familiar image:

A person at the top—deciding, directing, and guiding others through superior knowledge, experience, or judgment.

This image is deeply embedded in how we select leaders, train them, and evaluate them. We expect leaders to:

> know more
>
> see further
>
> decide better

But this expectation rests on an assumption that no longer holds:

That thinking happens inside the leader.

In a world of distributed cognition, this assumption collapses.

> This does not mean leadership disappears.
> It means leadership is no longer about individual cognition.

> Leaders are no longer the primary locus of intelligence.
> They are part of a system in which intelligence is shared across humans and machines.[1]

> This does not diminish leadership.
> It transforms it.

The End of the "Heroic Thinker"

The traditional model of leadership is built around the idea of the "heroic thinker":

the visionary who sees what others cannot

the expert who understands more deeply

the decision-maker who carries the cognitive burden

This model worked when:

information was limited

analysis was manual

cognition was internal

In that world, leadership advantage came from having more cognitive capacity than others.

But when AI systems can:

analyze vast datasets

generate strategies

simulate outcomes

no individual can maintain a decisive advantage based on internal cognition alone.

The leader is no longer the smartest entity in the room. The system is.[2]

From Decision-Maker to System Designer

If intelligence resides in the system, then leadership must shift accordingly.

From:

making decisions

To:

designing how decisions are made

This includes:

structuring human–AI interaction

defining roles within the system

establishing feedback loops

ensuring alignment with goals and values

Leaders become architects of cognition.[3]

They are responsible not just for outcomes, but for the processes that produce those outcomes.

Framing Becomes Power

In a distributed cognitive system, the most important decision is often the first one:

What is the problem?

AI systems can generate answers.
But they depend on how the question is framed.

A poorly framed problem leads to:

irrelevant outputs

misleading analyses

suboptimal decisions

A well-framed problem focuses the entire system.

In the age of AI, framing is power.[4]

Leaders must:

define objectives clearly

set constraints

articulate priorities

This is not a preliminary step.
It is the foundation of effective cognition.

Judgment Under Uncertainty

As AI expands the space of possibilities, leaders face a paradox:

more information

more options

more analysis

But also:

more uncertainty

more ambiguity

more complexity

AI does not eliminate uncertainty.
It often reveals more of it.

Leadership is no longer about having the right answer.
It is about making the right decision in the presence of many plausible answers.[5]

Maintaining Engagement

As AI systems become more capable, there is a risk that leaders disengage from the cognitive process.

They may:

defer too quickly to model outputs

rely on summaries without deeper examination

accept recommendations without sufficient challenge

This creates a new leadership failure mode:

passive decision-making

The danger is not that leaders will be replaced by AI. It is that they will stop thinking actively within the system.

Maintaining active judgment becomes critical.[6]

Designing for Alignment

In distributed systems, misalignment can occur at multiple levels:

between human intent and AI outputs

between different parts of the organization

between short-term actions and long-term goals

Leaders must ensure alignment by:

defining clear objectives

establishing consistent principles

creating mechanisms for feedback and correction

Alignment is not achieved once.
It must be maintained continuously.[7]

From Authority to Influence

In traditional hierarchies, leadership authority is tied to position.

In distributed cognitive systems, authority becomes less about control and more about influence.

Leaders influence:

how problems are framed

how systems are designed

how decisions are evaluated

Leadership shifts from commanding actions to shaping systems.[8]

Building Cognitive Capacity at Scale

A key responsibility of leadership is to enhance the cognitive capacity of the organization.

This includes:

integrating AI effectively

training individuals to work within human–AI systems

designing workflows that leverage distributed cognition

The goal is not to make individuals smarter in isolation.

It is to make the system more capable as a whole.[9]

The Ethics of Distributed Cognition

As cognition becomes distributed, ethical responsibility becomes more complex.

Leaders must consider:

how decisions are influenced by AI

how biases are introduced and propagated

how accountability is maintained

Ethics is no longer a separate function.
It is embedded in the design of cognitive systems.[10]

The New Leadership Profile

The effective leader in the cognitive era is characterized by:

System thinking

Cognitive design

Judgment

Adaptability

Alignment

This is a different kind of leadership.

It is less about individual brilliance.
More about systemic intelligence.

A New Definition of Leadership

Leadership is no longer the control of decisions.
It is the design of cognition.

We can redefine leadership:

Leadership is the design, orchestration, and continuous alignment of distributed cognitive systems to achieve collective goals under uncertainty.[11]

This includes:

framing problems

designing systems

guiding decisions

maintaining alignment

The Final Shift

At its core, the transformation of leadership reflects the same pattern we have seen throughout this book:

Cognition has moved.

From:

the individual

To:

the system

Leadership must move with it.

Transition

We have now traced the transformation across:

cognition

expertise

education

research

decision-making

governance

leadership

Each reveals the same underlying shift:

Intelligence is no longer contained.
It is distributed.[1][2]

This brings us to the final question:

What kind of future emerges when institutions are redesigned around this reality?

The next chapter looks forward—examining the trajectory of the cognitive revolution and the choices that will shape its outcome.

Not as prediction, but as design.

Endnotes

1. Leadership in this chapter is being redefined against the background already established by distributed cognition. Karl Weick's Sensemaking in Organizations (1995) is especially useful here because it treats leadership less as private brilliance and more as the shaping of collective interpretation under uncertainty.

2. On the limits of individual cognitive advantage, OpenAI's GPT-4 Technical Report (2023) and Dell'Acqua et al.'s 'Navigating the Jagged Technological Frontier' (2026) together make the point clearly: model systems can now outperform unaided professionals on some knowledge tasks, but they do so unevenly, which changes what leaders can plausibly dominate through personal expertise alone.

3. The argument that leaders become system designers is closer to Herbert Simon's The Sciences of the Artificial (1969) and Amershi et al.'s 'Guidelines for Human-AI Interaction' (2019) than to standard leadership manuals. The first frames organizations as designed systems; the second shows that outcomes in AI settings depend on the quality of those designs.

4. Problem framing has a long lineage in organizational thought, but it becomes operationally vivid in AI settings. Weick (1995) explains why sensemaking begins with how a situation is framed, and GPT-4's technical report (2023) makes obvious how dramatically outputs change when the framing changes.

5. For judgment under abundance, Amos Tversky and Daniel Kahneman's 'Judgment under Uncertainty' (1974) is still foundational, and Dell'Acqua et al. (2026) supplies the modern AI version: once many plausible outputs are cheaply available, the scarce capability is discriminating among them.

6. Over-reliance is best understood through the older automation literature. Raja Parasuraman and Victor Riley's 'Humans and Automation' (1997) and John Lee and Katrina See's 'Trust in Automation' (2004) show why people can surrender agency to systems that are fluent, fast, or opaque.

7. On alignment in distributed systems, Stuart Russell's Human Compatible (2019) is useful because it keeps the focus on goals and values rather than on raw capability, while NIST AI RMF (2023) translates that concern into practical governance language.

8. The shift from authority to influence is easier to grasp through Walter Powell's 'Neither Market nor Hierarchy' (1990) and Weick's Sensemaking in Organizations (1995) than through heroic-leadership writing. In networked systems, shaping flows of interpretation and coordination matters more than issuing orders from above.

9. Scaling cognitive capacity is one of the clearest empirical results of recent augmentation work. Brynjolfsson, Rock, and Syverson (2021) provides the complementarity logic, and Dell'Acqua et al. (2026) gives direct evidence that AI can raise output and quality when the task sits inside the model's frontier and the workflow is well matched.

10. Ethics in this chapter is not a separate chapter appended to leadership; it is part of system design. Russell (2019), NIST AI RMF (2023), and the OECD AI Principles (2019, revised 2024) all support that reading.

11. The chapter's redefinition of leadership is best read as a synthesis of Simon (1969), Weick (1995), and the distributed-cognition literature already introduced earlier. Leadership becomes the orchestration of a system that can think, not the display of a single mind's superiority.

12. The closing claim that no one thinks alone is simply the leadership version of the book's broader thesis. Hutchins's Cognition in the Wild (1995) and Clark and Chalmers's 'The Extended Mind' (1998) remain the shortest route back to that larger argument.

CHAPTER 13

The Cognitive Revolution

We often speak of artificial intelligence as a technological revolution.

It is not.

It is a cognitive revolution.[1]

What is changing is not just what machines can do, but where and how intelligence exists. The shift we are experiencing is not about tools becoming more powerful. It is about cognition moving beyond the boundaries that have defined it for centuries.

> This is not the rise of artificial intelligence.
> It is the reorganization of human intelligence.[2]

To understand what comes next, we must situate this moment within a broader historical arc.

Three Revolutions, Three Constraints

Human progress has been shaped by the removal of fundamental constraints:

Agricultural Revolution → removed biological scarcity (food)

Industrial Revolution → removed physical labor constraints

Cognitive Revolution → removes cognitive labor constraints[3]

Each transformation did more than increase productivity.
It redefined the structure of society.

Agriculture enabled settlement, cities, and civilizations.
Industry enabled scale, standardization, and global production.

Now, the cognitive revolution is beginning to reshape:

how we learn

how we discover

how we decide

how we organize

The binding constraint of the modern world is no longer food or labor.
It is cognition.

And that constraint is being lifted.

From Amplification to Transformation

Early narratives about AI focused on amplification:

faster analysis

better predictions

improved efficiency

But amplification is only the first phase.

As we have seen, AI does not just make existing processes more efficient. It changes their structure.

Nothing disappears.
It becomes abundant—and therefore ceases to define the system.

Learning becomes interaction

Research becomes evaluation

Decision-making becomes system-level

Leadership becomes cognitive design

AI does not simply accelerate cognition.
It transforms it.[4]

This transformation is deeper than most organizations—
and even most experts—fully recognize.

The Compression of Time

One of the most striking features of the current transition is
its speed.

The agricultural revolution unfolded over millennia.
The industrial revolution unfolded over centuries.

The cognitive revolution is unfolding over decades.[5]

This compression has consequences:

institutions have less time to adapt

individuals must re-skill continuously

advantages compound rapidly

The pace of change is no longer aligned with the pace of
institutional evolution.

This creates a widening gap between:

what is possible

and what is implemented

The Emergence of Cognitive Infrastructure

As AI becomes embedded in workflows, it begins to function as infrastructure.[6]

Like electricity or the internet, it becomes:

ubiquitous

invisible

essential

When this happens, the question shifts from:

"Should we use AI?"

To:

"How is cognition structured in our system?"

Organizations that treat AI as optional will fall behind. Those that integrate it into their core architecture will gain compounding advantages.

The future will not be defined by access to AI.
It will be defined by how deeply it is integrated into cognitive systems.

The Redistribution of Intelligence

As cognition becomes distributed, intelligence itself is redistributed.[7]

Across:

individuals

machines

networks

institutions

This redistribution has profound implications.

It lowers barriers to entry:

more people can perform complex tasks

capabilities become widely accessible

But it also raises the bar for differentiation:

basic competence becomes universal

advanced judgment becomes critical

When intelligence is distributed, advantage shifts from possession to orchestration.

Inequality in the Cognitive Era

The cognitive revolution will not affect everyone equally.[8]

New forms of inequality may emerge:

between those who can effectively work with AI and those who cannot

between organizations that redesign systems and those that do not

between regions that build cognitive infrastructure and those that lag

This is not just a technological divide.
It is a cognitive divide.

The gap will not be who has access to AI.
It will be who knows how to think with it.

The Risk of Misalignment

As AI becomes more embedded in cognition, the risks become more systemic.[9]

errors can propagate across systems

biases can be amplified

misaligned objectives can lead to unintended outcomes

These risks are not new.
But their scale and speed are.

Managing them requires:

better governance

stronger alignment mechanisms

continuous oversight

The power of distributed cognition must be matched by the discipline of its design.

The Window of Design

We are still early in the cognitive revolution.[10]

The systems we design now—educational models, research structures, governance frameworks—will shape its trajectory.

This creates a rare window:

to rethink assumptions

to redesign institutions

to define norms and principles

The future of cognition is not predetermined.
It is being designed.

The Choice

At a high level, institutions face a choice:[11]

Adapt superficially

adopt AI tools

preserve existing structures

move incrementally

Redesign fundamentally

reconfigure around distributed cognition

integrate AI as infrastructure

transform workflows and governance

The first path leads to:

>temporary gains

>long-term stagnation

>The second path leads to:

>sustained advantage

>structural transformation

>The difference is not adoption.
>It is architecture.

A New Cognitive Ecology

What emerges from this transformation is a new cognitive ecology:[1][2]

humans and AI interacting continuously

knowledge flowing across systems

decisions emerging from distributed processes

In this ecology:

>boundaries blur

>roles evolve

systems adapt

The individual remains important.
But no longer central in isolation.

Intelligence becomes an ecosystem property.

What Remains Human

Amid this transformation, an important question arises:

What remains uniquely human?

As AI takes on more cognitive tasks, human value shifts toward:

judgment

meaning-making

ethical reasoning

goal-setting

These are not replaced by AI.
They are amplified in importance.[13]

The more cognition is distributed, the more human judgment matters.

The Beginning, Not the End

It is tempting to view the current moment as a culmination.

It is not.

We are at the beginning of a longer transformation.

AI systems will continue to evolve

human–AI interaction will deepen

institutions will gradually adapt

What we are witnessing is not a final state, but a transition.

A Final Reframe

We can now restate the central argument of this book:

> Intelligence has never been confined to the human brain. It has always been extended through artifacts and systems.

Artificial intelligence is the latest and most powerful extension—one that actively participates in cognition.

> As a result, intelligence is no longer an individual property.
> It is a distributed system phenomenon.[14]

> This is the cognitive revolution.

Closing Transition

The implications of this shift are vast.

They touch:

> how we think

> how we learn

> how we work

> how we lead

> how we organize

But the most important implication is this:

The future will not be shaped by those who simply use AI.
It will be shaped by those who design systems in which intelligence can emerge.

This is the challenge—and the opportunity—of our time.

The work now is not to understand the cognitive revolution.

It is to build for it.

Endnotes

1. Calling this an AI revolution is descriptively true but conceptually incomplete. Merlin Donald's Origins of the Modern Mind (1991) and Edwin Hutchins's Cognition in the Wild (1995) help make the stronger point: what is changing is the location and organization of cognition, not only the power of a new tool.

2. The reorganization of intelligence is already the thread running from the Preface through Chapter 12. The closest sources remain Hutchins (1995), Zhang's 'The Nature of External Representations in Problem Solving' (1997), and Hollan, Hutchins, and Kirsh (2000), all of which push intelligence outward from the individual to the system.

3. For the claim that major revolutions remove binding constraints, Joel Mokyr's The Lever of Riches (1990) is helpful as a big-picture historical frame, and Bresnahan and Trajtenberg's 'General Purpose Technologies "Engines of Growth"?' (1995) supplies the economic mechanism by which such shifts propagate.

4. The difference between amplification and transformation is exactly the difference between a helpful tool and a reorganized system. GPT-4 (2023), Brown et al. (2020), and Amershi et al. (2019) together make that transition visible.

5. The time-compression point is less a precise law than a historical comparison, but Brynjolfsson, Rock, and Syverson's 'The Productivity J-Curve' (2021) is still useful because it shows how digital general-purpose technologies diffuse quickly at the surface while institutional complements lag behind.

6. On AI as infrastructure, the core sources remain Bresnahan and Trajtenberg (1995), David (1990), and Brynjolfsson et al. (2021). Once a technology becomes infrastructural, the main question becomes how cognition is organized around it.

7. The redistribution of intelligence is simply distributed cognition scaled through modern systems. Hutchins (1995) and Hollan,

Hutchins, and Kirsh (2000) give the conceptual frame; Dell'Acqua et al. (2026) shows how redistributed capability behaves inside knowledge work.

8. The coming inequality is unlikely to be explained by raw access alone. OECD's Governing with Artificial Intelligence (2025) points to widespread skills and implementation gaps, and Dell'Acqua et al. (2026) suggests that advantage increasingly depends on whether people and institutions know how to work inside the model's frontier.

9. On systemic risk, NIST AI RMF (2023) and the OECD AI Principles (2019, revised 2024) are the right sources because both treat risk as something that scales with integration, context, and governance rather than as a property of model error alone.

10. The idea of a design window is closest to path-dependence work. Paul David's 'Clio and the Economics of QWERTY' (1985) and Douglass North's Institutions, Institutional Change and Economic Performance (1990) both show how early choices can harden into long-lived institutional patterns.

11. The institutional choice between superficial adaptation and deep redesign is the same choice already visible in Orlikowski (2000) and Brynjolfsson et al. (2021). Architecture, not rhetoric, determines the outcome.

12. A new cognitive ecology is exactly the sort of thing Donald (1991), Hutchins (1995), and Clark and Chalmers (1998) help describe: cognition dispersed across brains, artifacts, and social systems rather than enclosed within a single agent.

13. On what remains human, Stuart Russell's Human Compatible (2019) is useful because it keeps goals and values at the center, while McDuff et al. (2025) is useful because it shows, in practice, that human judgment still matters most when AI expands the space of plausible options.

14. The final claim that intelligence is now a distributed system phenomenon is not a new note so much as a synthesis of the book's preceding ones. Hutchins (1995), Zhang and Norman (1994), Zhang (1997), and Hollan et al. (2000) are still the most direct foundations.

CHAPTER 14

Designing for a World That Thinks

Understanding the cognitive revolution is not enough.

We can describe how cognition has moved.
We can analyze how institutions are misaligned.
We can diagnose innovation theater and redefine expertise.

But none of this changes reality unless it leads to design.

The central task of this era is not adoption.
It is construction.[1]

We are not passive observers of the cognitive revolution.
We are its architects.

From Description to Design

Most discussions of AI remain descriptive:

what the technology can do

how it affects jobs

what risks it introduces

These are important. But they stop short of the real challenge.

The real challenge is:

How do we design systems in which distributed cognition functions effectively?[2]

How do we organize human–AI interaction at scale?

How do we build institutions that learn, decide, and adapt continuously?

The question is no longer "What is AI doing?"
It is "What should we build?"

Designing Cognitive Systems

Design in the cognitive era is not primarily about products, interfaces, or services.

It is about systems of thinking.[3]

A cognitive system includes:

humans

AI models

data

workflows

feedback mechanisms

These elements must be intentionally structured.

If they are not, the system will still function—but often poorly:

misaligned outputs

inefficient processes

fragmented decision-making

Distributed cognition is inevitable.
Effective distributed cognition is designed.

The Principles of Cognitive Design

To move from analysis to construction, we need a set of design principles.

1. Design for Interaction, Not Isolation

Cognition now occurs in interaction loops between humans and AI.

Design must focus on:

> how questions are framed

> how outputs are presented

> how iteration occurs

> The quality of the system depends on the quality of interaction.[4]

2. Make Cognition Visible

In distributed systems, thinking is partially external.

Design should:

> expose intermediate steps

> make reasoning interpretable

> allow users to trace and question outputs

> Invisible cognition leads to:

> over-trust

> misuse

> error propagation

> Design must make cognition visible.[5]

3. Align Roles Within the System

Humans and AI have different strengths.

Design must clarify:

> what the human is responsible for

what the AI contributes

how responsibilities shift across contexts

Ambiguity leads to:

duplication

gaps

failure

Alignment is a design problem.[6]

4. Build Feedback Loops

Cognitive systems must learn continuously.

Design should include:

mechanisms for correction

integration of outcomes

iterative refinement

Without feedback, systems stagnate.[7]

5. Optimize for Judgment, Not Just Output

Outputs are easy to generate.
Judgment is harder.

Design should support:

evaluation

comparison

decision-making

The goal is not more answers.
It is better decisions.[8]

6. Integrate Across Boundaries

Cognition is no longer confined to disciplines or departments.

Design must:

connect data sources

enable cross-domain reasoning

support integration

Fragmentation reduces system intelligence.[9]

From Workflow to Cognitive Flow

Traditional workflows are:

linear

segmented

task-oriented

Cognitive systems require flows that are:

iterative

integrated

adaptive

Instead of:

step 1 → step 2 → step 3

We have:

frame → generate → evaluate → refine → decide → learn[10]

This cycle repeats continuously.

The unit of work becomes the cognitive loop.

Designing Roles and Capabilities

As systems change, roles must be redesigned.

Key capabilities include:

Framing — defining problems

Interaction — working effectively with AI

Evaluation — assessing outputs

Integration — combining insights

Decision-making — acting under uncertainty

Roles are no longer defined by tasks, but by contributions to the cognitive system.[11]

Infrastructure as Design

Design is not only conceptual.
It requires infrastructure.

Cognitive systems depend on:

data platforms

AI integration

interoperability

real-time access

Without infrastructure, design remains theoretical.

Infrastructure is the substrate of cognition.[12]

Designing for Trust

Trust becomes central in human–AI systems.

Users must trust:

the outputs

the system

the interaction

But trust cannot be assumed.
It must be designed.

This includes:

transparency

consistency

clear limitations

mechanisms for challenge and override

Trust is not a feature.
It is an outcome of design.[13]

The Cost of Poor Design

If cognitive systems are not well designed, the consequences are significant:

poor decisions

inefficient workflows

over-reliance or under-utilization of AI

loss of trust

These are not technical failures.
They are design failures.[14]

The Emergence of Cognitive Architecture

As design becomes central, a new discipline emerges:

Cognitive architecture at the institutional level

This includes:

structuring learning systems

designing research processes

building decision frameworks

creating governance models

It is not confined to a single role or department.

It is a leadership responsibility.[15]

Designing at Multiple Levels

Cognitive systems must be designed at multiple levels:

Individual level
Skills for interacting with AI

Team level
Collaborative workflows

Organizational level
Integrated systems and structures

Institutional level
Governance, incentives, and culture

Alignment across these levels is critical.[16]

From Users to Participants

In traditional systems, people are users of tools.

In cognitive systems, they are participants in thinking processes.

This changes expectations:

from passive consumption

to active engagement

Design must support participation, not just usage.[17]

The Design Imperative

We can now state the imperative clearly:

If cognition is distributed, it must be designed.[18]

Otherwise:

systems will be inefficient

decisions will be suboptimal

potential will be unrealized

Design is not optional.
It is the core work of the cognitive era.

A New Role: The Cognitive Architect

This leads to the emergence of a new role:

The cognitive architect

This role is responsible for:

designing human–AI systems

structuring workflows

aligning components

ensuring continuous learning

It may be embodied in:

leaders

teams

institutions

But the function must exist.[19]

Closing Reflection

We have moved from:

understanding cognition

to recognizing its redistribution

to redefining institutions

to designing systems

This is the arc of the book.

The final step is to bring it together.

Not as theory, but as a coherent vision:

What does a fully realized AI-native world look like?
What principles guide it?
What choices shape it?

The final chapter offers that synthesis.

Not as a prediction of what will happen.

But as a statement of what should be built.

Endnotes

1. If design becomes the central task, Herbert Simon's The Sciences of the Artificial (1969) is still the unavoidable starting point. Wanda Orlikowski's 'Using Technology and Constituting Structures' (2000) shows why that design question cannot be separated from practice.

2. Designing distributed cognition requires more than acknowledging it. Hutchins's Cognition in the Wild (1995) and Zhang's 'The Nature of External Representations in Problem Solving' (1997) are useful because they specify what has to be designed: representations, roles, tools, and flows.

3. On systems of thinking, Hollan, Hutchins, and Kirsh's 'Distributed Cognition' (2000) and Simon (1969) fit together well.

One gives the cognitive architecture; the other gives the design vocabulary.

4. The most practical interaction-design reference remains Amershi et al.'s 'Guidelines for Human-AI Interaction' (2019). It is especially relevant to this chapter because it translates broad ideas about human-AI systems into concrete design principles around expectations, timing, explanation, and recovery.

5. Visibility matters because opaque systems distort trust. Joshua Kroll et al.'s 'Accountable Algorithms' (2017) and Lee and See's 'Trust in Automation' (2004) both explain why users need enough visibility into a system's operation to calibrate reliance rather than drift into over-trust or under-trust.

6. Role alignment is one of the oldest and most persistent design questions in automation. Parasuraman and Riley (1997) and Amershi et al. (2019) are the clearest pair here because they explain both the risks of poor task allocation and the interaction patterns that mitigate them.

7. Feedback loops belong both to learning theory and to AI-mediated work. Argyris and Schon (1978) and Senge (1990) provide the organizational frame; Nathan J. Szymanski et al.'s A-Lab paper (2023) shows what a tightly coupled propose-test-refine loop looks like in practice.

8. On judgment, Amos Tversky and Daniel Kahneman (1974) remains the classic warning that decision quality is not guaranteed by more information alone. Simon (1969) complements that by reminding us that design shapes the environment in which judgment occurs.

9. Integration across boundaries is partly a cognitive issue and partly an infrastructural one. Hutchins (1995) gives the cognitive reason, while Wilkinson et al.'s FAIR principles (2016) give the infrastructural conditions under which data and models can actually interoperate across domains.

10. The chapter's cognitive loop - frame, generate, evaluate, refine, decide, learn - is no longer just a metaphor. A-Lab (2023) and McDuff et al. (2025) both show iterative human-machine loops in operation, one in experimental science and the other in diagnostic reasoning.

11. On role transformation, the most relevant recent evidence is Dell'Acqua et al. (2026), which shows that AI changes the content of work unevenly rather than uniformly. That is why this chapter redefines roles around framing, evaluation, and orchestration.

12. Infrastructure is not background plumbing here; it is part of cognition. Bresnahan and Trajtenberg (1995) explains why infrastructure changes whole systems, and Wilkinson et al. (2016) explains why data and metadata standards are part of what makes cognitive infrastructure usable.

13. Trust is designed, not assumed. Lee and See (2004) remains the classic treatment of calibrated reliance, and NIST AI RMF (2023) updates that concern in the language of trustworthy AI practice.

14. Design failure often looks like technical failure from the outside. Parasuraman and Riley (1997) and Orlikowski (2000) both help explain why misaligned workflows, poor role allocation, or badly timed automation can degrade performance even when the underlying tool is capable.

15. The chapter's phrase 'cognitive architecture' is meant in a system-design sense, not merely a computational one. Simon (1969) and Hutchins (1995) are the two most important touchstones for that usage.

16. Multi-level alignment is a systems problem. Senge's The Fifth Discipline (1990) and Douglass North's Institutions (1990) are useful together because one emphasizes interconnected learning processes and the other emphasizes how rules and incentives structure behavior over time.

17. Participation is a key shift in this book's overall argument. Amershi et al. (2019) makes the point operationally for human-AI interfaces, while Hutchins (1995) makes it conceptually for distributed cognition more broadly.

18. The design imperative itself is really just the practical conclusion of the earlier chapters. Simon (1969), Orlikowski (2000), and the distributed-cognition literature together explain why undesigned systems do not stay neutral; they drift toward whatever their inherited structures permit.

19. 'Cognitive architect' is not yet a settled institutional title, but the function is already implicit in Simon's design tradition (1969),

Amershi et al.'s interaction guidance (2019), and the workflow redesign literature used throughout this book. The point is not the label; it is that someone must own the design of the human-AI cognitive system.

CHAPTER 15

The World We Choose to Build

Every revolution creates two futures.

One emerges by default—through inertia, partial adaptation, and the accumulation of small, uncoordinated decisions.

The other is constructed—through deliberate design, clear principles, and aligned action.

The cognitive revolution will be no different.

> The future of intelligence is not predetermined.[1]
> It will be shaped by what we choose to build.

The Default Future

If we do nothing—if we continue along current trajectories—an implicit future begins to take shape.[2]

In this future:

AI is widely deployed but unevenly integrated

institutions layer new capabilities onto old structures

decision-making becomes faster, but not necessarily better

expertise becomes more superficial in many contexts

inequality grows between those who can and cannot navigate distributed cognition

Innovation continues, but largely at the surface.

Organizations become:

more complex

more fragmented

more dependent on systems they do not fully understand

Intelligence expands, but coherence declines.[3]

This is the path of least resistance.

It does not require intentional design.
It emerges from incremental adaptation.

The Designed Future

There is another path.

One in which institutions recognize that cognition has changed—and redesign themselves accordingly.

In this future:

AI is integrated as infrastructure, not layered as a tool

human–AI systems are intentionally designed

learning, discovery, and decision-making operate as continuous loops

governance evolves to guide distributed cognition effectively

leadership focuses on alignment, judgment, and system design

Organizations become:

more adaptive

more coherent

more capable of learning and evolving

Intelligence is not just expanded.
It is organized.[4]

This future does not emerge automatically.
It must be built.

Principles for the Cognitive Era

To build effectively, we need guiding principles.

These are not technical rules.
They are design commitments.

1. Intelligence Is a System Property

Stop optimizing individuals in isolation.
Start designing systems in which intelligence emerges.[5]

2. AI Is Infrastructure

Treat AI as foundational, not optional.
Design workflows, roles, and decisions around it.[6]

3. Judgment Is the Core Human Capability

Shift focus from knowledge acquisition to:

evaluation

decision-making

ethical reasoning

Judgment anchors outcomes in distributed systems.[7]

4. Learning Must Be Continuous

Eliminate rigid boundaries between:

learning and work

training and practice

Design systems that learn as they operate.[8]

5. Governance Must Enable, Not Constrain

Replace rigid control with:

adaptive guardrails

continuous oversight

alignment mechanisms

Governance should shape cognition, not slow it.[9]

6. Design Is the Central Discipline

Recognize that outcomes depend on:

how systems are structured

how interactions are designed

how components are aligned

Design determines system intelligence.[10]

The Role of Institutions

Institutions play a critical role in shaping the future of cognition.

They:

organize learning

structure research

guide decision-making

establish norms and values

If institutions fail to evolve:

the cognitive revolution will be fragmented

benefits will be unevenly distributed

risks will be harder to manage

If institutions redesign themselves:

intelligence can be scaled effectively

systems can remain aligned

progress can be sustained

Institutions are the primary vehicles through which cognition is organized at scale.[11]

The Human Role

Amid this transformation, it is important to reaffirm the role of humans.

AI expands cognitive capability.
It does not define purpose.

Humans remain responsible for:

defining goals

setting values

making ethical decisions

guiding systems

The more intelligence is distributed, the more responsibility becomes human.[12]

Avoiding Determinism

There is a tendency to view technological change as inevitable and deterministic.

This perspective is misleading.

Technologies create possibilities.
They do not dictate outcomes.

How AI is used, integrated, and governed depends on human choices.

The cognitive revolution is not something happening to us.
It is something we are creating.[13]

The Work Ahead

Building AI-native institutions requires:

rethinking assumptions

redesigning structures

developing new capabilities

aligning systems with values

It is not:

a single initiative

a short-term project

a technical upgrade

It is an ongoing process.[14]

A Moment of Responsibility

We are at an inflection point.

The systems we design now will:

shape how intelligence is distributed

influence how decisions are made

determine how knowledge evolves

This is a moment of responsibility as much as opportunity.[15]

A Final Reframe

We began with a simple idea:

Intelligence is no longer confined to the brain.

We have followed its implications across:

cognition

expertise

education

research

decision-making

governance

leadership

institutional design

The conclusion is clear:

The future belongs to systems that can think.[16]

But these systems do not emerge on their own.

They are designed.

Closing

The cognitive revolution is not defined by the rise of artificial intelligence.

It is defined by the relocation of intelligence.

From:

individuals

To:

systems

This shift changes everything.[17]

It challenges our assumptions about:

knowledge

expertise

learning

leadership

It requires us to redesign the institutions that shape our world.

And it offers a choice:

To adapt incrementally and remain constrained by legacy structures.

Or to design intentionally—and build systems in which intelligence can truly emerge.

The question is not whether the world will change. It is whether we will design it well.[18]

End of Book

Endnotes

1. This chapter's contrast between default and designed futures is best read against Douglass North's Institutions, Institutional Change and Economic Performance (1990) and Langdon Winner's 'Do Artifacts Have Politics?' (1980). Technologies open possibilities, but institutions and design choices determine which possibilities harden into social reality.

2. The default future is the path-dependence story. Paul David's 'Clio and the Economics of QWERTY' (1985), David's 'The Dynamo and the Computer' (1990), and North (1990) all show how systems drift forward by inheritance when they are not deliberately redesigned.

3. On complexity and fragmentation, Herbert Simon's 'The Architecture of Complexity' (1962) remains a classic. It is useful here because it explains why uncoordinated growth produces brittle systems even when each local addition looks sensible on its own.

4. The designed future described in this chapter follows directly from Simon's The Sciences of the Artificial (1969) and Orlikowski's 'Using Technology and Constituting Structures' (2000). Systems become coherent when they are intentionally structured around how work and cognition actually happen.

5. The principle that intelligence is a system property has already been established earlier, but the clearest sources remain Hutchins's Cognition in the Wild (1995) and Zhang's 'The Nature of External Representations in Problem Solving' (1997). This chapter is drawing the institutional consequence of that claim.

6. The principle that AI is infrastructure is grounded in Bresnahan and Trajtenberg (1995), David (1990), and Brynjolfsson, Rock, and Syverson (2021). The strategic implication is the same one repeated across the book: infrastructure changes the design problem.

7. Judgment becomes central whenever option sets expand faster than certainty does. Tversky and Kahneman's 'Judgment under Uncertainty' (1974) remains the classic frame, and McDuff et al. (2025) shows the same issue in a modern interactive AI setting.

8. On continuous learning, Argyris and Schon (1978) and Senge (1990) are still the strongest organizational anchors. This principle matters here because AI-native systems cannot rely on episodic retraining alone; they must learn in operation.

9. The governance principle in this chapter is well captured by NIST AI RMF (2023) and the OECD AI Principles (2019, revised 2024). Both assume that effective governance enables action inside structured constraints rather than halting action altogether.

10. Design as discipline is a straight line back to Simon (1969) and, in the human-AI setting, to Amershi et al. (2019). It is included here because the chapter's normative argument depends on design being treated as a core capability, not as an afterthought.

11. Institutions at scale are the place where cognition, rules, and incentives meet. North (1990) remains the most useful single reference for that level of analysis.

12. On human responsibility, Stuart Russell's Human Compatible (2019) and the OECD AI Principles (2019, revised 2024) are both helpful because they insist that goals, values, and acceptable trade-offs remain human responsibilities even when systems become more capable.

13. The anti-determinist point is important. Winner's 'Do Artifacts Have Politics?' (1980) and Orlikowski (2000) both help show why technologies do not arrive with a single social destiny attached to them.

14. Transformation as process is already implicit in Argyris and Schon (1978) and is made economically concrete in Brynjolfsson, Rock, and Syverson's 'The Productivity J-Curve' (2021). The main lesson is that redesign unfolds through iterative institutional change, not one-off announcement.

15. On inflection points, David's QWERTY paper (1985) and North (1990) are again the right references because they show how early design choices can become durable constraints or advantages later.

16. The claim that the future belongs to systems that can think is the end-state version of Herbert Simon's systems view and Hutchins's distributed-cognition view. Simon's 'The Architecture of Complexity' (1962) and Hutchins's Cognition in the Wild (1995) make that combination intelligible.

17. The relocation-of-intelligence claim can now be cited more briefly because it has been developed throughout the manuscript. Hutchins (1995), Brown et al. (2020), and GPT-4 (2023) are enough to anchor it here.

18. The closing claim about design responsibility is the point where policy and cognition meet. NIST AI RMF (2023), the OECD AI Principles (2019, revised 2024), and Winner (1980) all support

the idea that outcomes depend not just on capability, but on deliberate choices about how systems are built and governed.

The Brain Is Now Open Source

This is not the end of the book.

It is the beginning of a different way to understand intelligence.

We began with a question:

Where does intelligence reside?

For centuries, the answer seemed obvious.

In the human mind.

Everything we built—education, professions, institutions—rested on this assumption. Intelligence was cultivated within individuals, measured within individuals, and rewarded within individuals.

Even when we collaborated, we believed thinking ultimately belonged to the person.

That belief has now broken.[1]

Not suddenly.
Not completely.
But irreversibly.

The boundary of intelligence has shifted.

From Contained to Connected

Across history, cognition has expanded beyond the brain.

Language allowed us to think together.
Writing allowed us to remember beyond ourselves.
Mathematics allowed us to reason in structured systems.

Each step moved thinking outward—into shared spaces, external representations, and collective processes.[2]

But the human mind remained the center.

The artifact supported.
The human thought.

That distinction no longer holds.

Artificial intelligence changes the relationship.[3]

The artifact does not just store or structure thought.
It participates in it.

It generates.
It evaluates.
It responds.

For the first time, cognition is not only extended.
It is shared with something that thinks back.

The Dissolution of the Boundary

We still speak as if thinking happens inside the head.

But in practice, it increasingly happens elsewhere:

in prompts and responses

in models and data

in interactions that unfold across systems

The mind is no longer a closed space.
It is part of a network.[4]

Intelligence has moved from containment to connection.

This does not mean the human disappears.

It means the human is no longer the sole locus of cognition.

The brain becomes:

a participant

a coordinator

a source of judgment

But not the entire system.

What It Means to Be Intelligent

If intelligence is no longer contained within individuals, then the definition of intelligence must change.

It is no longer:

what you know

what you can recall

what you can produce alone

It becomes:

how you engage with systems

how you structure interactions

how you evaluate and decide

Intelligence is no longer an attribute.
It is a relationship.[5]

A relationship between:

human and machine

question and response

system and context

The Irreversibility

This shift will not reverse.

We will not return to a world where:

knowledge must be internalized to be useful

reasoning occurs only within the individual

cognition is bounded by biology

The trajectory is clear:

systems will become more capable

interactions will become more seamless

cognition will become more distributed

Once intelligence leaves the boundary of the brain, it does not return.[6]

Institutions that resist this shift will struggle.

Individuals who ignore it will fall behind.

Not because they lack intelligence, but because they are operating within an outdated model of it.

The New Responsibility

If intelligence is now distributed, then its design becomes a responsibility.[7]

Not just for engineers.
Not just for technologists.

For leaders.
For educators.
For institutions.

Because distributed cognition does not organize itself optimally.

Without design, it becomes:

fragmented

misaligned

inefficient

The question is no longer whether intelligence is distributed.
It is whether it is designed well.

The Risk

There is a risk in this transformation.

If poorly designed:

systems can amplify error

biases can propagate

decision-making can degrade

If unaligned:

cognition can become fragmented

responsibility can become unclear

trust can erode

The power of distributed intelligence is matched by its fragility.[8]

The Opportunity

But there is also opportunity.

If well designed:

cognition can scale beyond individual limits

learning can become continuous

decisions can improve

discovery can accelerate

We can build systems that:

think more effectively

adapt more rapidly

align more closely with human values

Intelligence can become a shared, scalable resource.[9]

A New Mental Model

To move forward, we need a new mental model.

Not:

the mind as a container

But:

cognition as a system

Not:

intelligence as possession

But:

intelligence as participation

Not:

thinking as internal

But:

thinking as interaction[10]

The Final Shift

At the deepest level, the transformation is simple:

The brain is no longer the boundary of intelligence.[11]

It is one component in a larger system.

A powerful component.
An essential component.

But no longer the whole.

The Line

We can now say it clearly:

The brain is now open source.

Not because it is exposed.
But because intelligence is no longer exclusive.

It is shared.
It is distributed.
It is accessible.

It can be engaged by anyone, anywhere, often at near-zero cost.[12]

Intelligence is no longer something you possess.
It is something you access.

What Comes Next

What comes next is not determined by technology alone.

It depends on how we respond:

whether we redesign institutions

whether we rethink education

whether we build effective cognitive systems

whether we align intelligence with human values

The future is not:

human vs machine

It is:

human with machine

And more importantly:

how that system is designed[13]

Closing

We began with the belief that intelligence resides in the individual; we end with a clearer truth: intelligence emerges from systems.[14] This is the cognitive revolution—not the rise of machines, but the reorganization of thinking. Intelligence has shifted from private to shared, from contained to distributed, from owned to accessed, from scarce to universally available. The boundary has moved. The system has expanded. The responsibility is now ours: not to adapt to intelligence, but to design it.

Endnotes

1. The claim that the old picture of intelligence has broken is the final restatement of the argument introduced in the Preface and Chapter 1. Hutchins's Cognition in the Wild (1995) and Clark and Chalmers's 'The Extended Mind' (1998) remain the clearest sources for why intelligence cannot be treated as wholly sealed inside the individual mind.

2. For the long outward expansion of cognition, the key historical sources are Merlin Donald's Origins of the Modern Mind (1991), Walter Ong's Orality and Literacy (1982), and Jack Goody's The Logic of Writing and the Organization of Society (1986). Together they show that language, writing, and formal systems progressively relocate cognitive work into shared external media.

3. What AI changes is not the existence of cognitive extension but the activity of the artifact. Brown et al.'s 'Language Models are Few-Shot Learners' (2020), OpenAI's GPT-4 Technical Report (2023), and Boiko et al.'s 'Autonomous Chemical Research with Large Language Models' (2023) all show systems that generate, respond, and help carry a reasoning process forward.

4. The claim that the mind is now part of a network is most directly supported by Clark and Chalmers's 'The Extended Mind' (1998), Hollan, Hutchins, and Kirsh's 'Distributed Cognition' (2000), and Zhang's 'The Nature of External Representations in Problem Solving' (1997). The point is not metaphorical; it is functional.

5. On intelligence as relationship rather than possession, Hutchins (1995), Zhang and Norman (1994), and Zhang (1997) remain the closest sources. They all, in different ways, define intelligence through coordinated interaction with representations, artifacts, and other actors.

6. The irreversibility claim is better read historically than dramatically. Donald (1991) and Andy Clark's Natural-Born Cyborgs (2003) both suggest that once cognition is successfully reorganized through external supports, human intelligence does not simply retreat back to its prior boundary.

7. If intelligence is distributed, design responsibility follows. Herbert Simon's The Sciences of the Artificial (1969), Amershi et al.'s 'Guidelines for Human-AI Interaction' (2019), and NIST AI RMF (2023) are useful together because they show, respectively, why systems are designed, how human-AI interaction has to be designed, and how such systems have to be governed.

8. The fragility of distributed intelligence is the familiar problem of miscalibrated automation in a stronger form. Parasuraman and Riley (1997), Lee and See (2004), and NIST AI RMF (2023) all support the idea that capability without calibrated use can amplify error rather than reduce it.

9. The opportunity side is just as real. McDuff et al.'s 'Towards Accurate Differential Diagnosis with Large Language Models' (2025), Dell'Acqua et al.'s 'Navigating the Jagged Technological Frontier' (2026), and Boiko et al. (2023) all show cases in which well-designed human-AI systems expand cognitive reach and improve performance.

10. The new mental model advanced here - from container to system, from possession to participation - is essentially a synthesis of Hutchins (1995), Clark and Chalmers (1998), and Zhang (1997). By this point in the book, those sources should feel less like supporting citations and more like the conceptual spine.

11. 'The brain is no longer the boundary of intelligence' is the book's distilled proposition. Clark and Chalmers (1998), Hutchins (1995), and Clark's Natural-Born Cyborgs (2003) are the shortest path back to it.

12. The near-zero-cost-access claim is tied to the public scaling of large language models. Brown et al. (2020) marks the capability inflection, and GPT-4 (2023) marks the broadening of sophisticated language-based reasoning as an accessible interface rather than a specialist research artifact.

13. The future-depends-on-design claim returns us to Simon (1969), Amershi et al. (2019), and NIST AI RMF (2023). The technical system matters, but so do the interaction rules, the institutions, and the values embedded in them.

14. The closing sentence is simply the full synthesis of the manuscript. Hutchins (1995), Zhang and Norman (1994), Zhang (1997), and Hollan et al. (2000) together justify the move from intelligence as a private possession to intelligence as an emergent property of systems.

Bibliography

Amershi, Saleema, Dan Weld, Mihaela Vorvoreanu, Andrew Fourney, Besmira Nushi, Penny Collisson, Jina Suh, et al. "Guidelines for Human-AI Interaction." In *Proceedings of the 2019 CHI Conference on Human Factors in Computing Systems*, 1–13. New York: Association for Computing Machinery, 2019. https://doi.org/10.1145/3290605.3300233.

Argyris, Chris, and Donald A. Schön. *Organizational Learning: A Theory of Action Perspective*. Reading, MA: Addison-Wesley, 1978.

Barocas, Solon, and Andrew D. Selbst. "Big Data's Disparate Impact." *California Law Review* 104, no. 3 (2016): 671–732. https://doi.org/10.15779/Z38BG31.

Bender, Emily M., Timnit Gebru, Angelina McMillan-Major, and Shmargaret Shmitchell. "On the Dangers of Stochastic Parrots: Can Language Models Be Too Big?" In *Proceedings of the 2021 ACM Conference on Fairness, Accountability, and Transparency*, 610–623. New York: Association for Computing Machinery, 2021. https://doi.org/10.1145/3442188.3445922.

Boiko, Daniil A., Robert MacKnight, Ben Kline, and Gabe Gomes. "Autonomous Chemical Research with Large Language Models." *Nature* 624 (2023): 570–578. https://doi.org/10.1038/s41586-023-06792-0.

Bran, Andres M., Sam Cox, Oliver Schilter, Carlo Baldassari, Andrew D. White, and Philippe Schwaller. "Augmenting Large Language Models with Chemistry Tools." *Nature*

Machine Intelligence 6 (2024): 525–535.
https://doi.org/10.1038/s42256-024-00832-8.

Bresnahan, Timothy F., and Manuel Trajtenberg. "General Purpose Technologies 'Engines of Growth'?" *Journal of Econometrics* 65, no. 1 (1995): 83–108.
https://doi.org/10.1016/0304-4076(94)01598-T.

Brown, John Seely, Allan Collins, and Paul Duguid. "Situated Cognition and the Culture of Learning." *Educational Researcher* 18, no. 1 (1989): 32–42.
https://doi.org/10.3102/0013189X018001032.

Brown, Tom B., Benjamin Mann, Nick Ryder, Melanie Subbiah, Jared D. Kaplan, Prafulla Dhariwal, Arvind Neelakantan, et al. "Language Models are Few-Shot Learners." In *Advances in Neural Information Processing Systems* 33 (2020): 1877–1901.

Brynjolfsson, Erik, Daniel Rock, and Chad Syverson. "The Productivity J-Curve: How Intangibles Complement General Purpose Technologies." *American Economic Journal: Macroeconomics* 13, no. 1 (2021): 333–372.
https://doi.org/10.1257/mac.20180386.

Clark, Andy. *Natural-Born Cyborgs: Minds, Technologies, and the Future of Human Intelligence.* Oxford: Oxford University Press, 2003.

Clark, Andy, and David J. Chalmers. "The Extended Mind." *Analysis* 58, no. 1 (1998): 7–19.
https://doi.org/10.1093/analys/58.1.7.

Collins, Harry, and Robert Evans. *Rethinking Expertise.* Chicago: University of Chicago Press, 2007.

David, Paul A. "Clio and the Economics of QWERTY." *American Economic Review* 75, no. 2 (1985): 332–337.

David, Paul A. "The Dynamo and the Computer: An Historical Perspective on the Modern Productivity Paradox." *American Economic Review* 80, no. 2 (1990): 355–361.

Dell'Acqua, Fabrizio, Edward McFowland III, Ethan Mollick, Hila Lifshitz-Assaf, Katherine C. Kellogg, Saran Rajendran, Lisa Krayer, François Candelon, and Karim R. Lakhani. "Navigating the Jagged Technological Frontier: Field Experimental Evidence of the Effects of Artificial Intelligence on Knowledge Worker Productivity and Quality." *Organization Science* 37, no. 2 (2026): 403–423. https://doi.org/10.1287/orsc.2025.21838.

Donald, Merlin. *Origins of the Modern Mind: Three Stages in the Evolution of Culture and Cognition.* Cambridge, MA: Harvard University Press, 1991.

Ericsson, K. Anders, Neil Charness, Paul J. Feltovich, and Robert R. Hoffman, eds. *The Cambridge Handbook of Expertise and Expert Performance.* Cambridge: Cambridge University Press, 2006.

Gibson, James J. *The Ecological Approach to Visual Perception.* Boston: Houghton Mifflin, 1979.

Goody, Jack. *The Logic of Writing and the Organization of Society.* Cambridge: Cambridge University Press, 1986.

Gottweis, Juraj, Wei-Hung Weng, Alexander Daryin, Tao Tu, Anil Palepu, Petar Sirkovic, and Artiom Myaskovsky, et al. "Towards an AI Co-Scientist." arXiv, 2025. https://doi.org/10.48550/arXiv.2502.18864.

Hollan, James, Edwin Hutchins, and David Kirsh. "Distributed Cognition: Toward a New Foundation for Human-Computer Interaction Research." *ACM*

Transactions on Computer-Human Interaction 7, no. 2 (2000): 174–196. https://doi.org/10.1145/353485.353487.

Hutchins, Edwin. *Cognition in the Wild*. Cambridge, MA: MIT Press, 1995.

Jumper, John, Richard Evans, Alexander Pritzel, Tim Green, Michael Figurnov, Olaf Ronneberger, Kathryn Tunyasuvunakool, et al. "Highly Accurate Protein Structure Prediction with AlphaFold." *Nature* 596 (2021): 583–589. https://doi.org/10.1038/s41586-021-03819-2.

Krizhevsky, Alex, Ilya Sutskever, and Geoffrey E. Hinton. "ImageNet Classification with Deep Convolutional Neural Networks." In *Advances in Neural Information Processing Systems* 25 (2012): 1097–1105.

Kroll, Joshua A., Joanna Huey, Solon Barocas, Edward W. Felten, Joel R. Reidenberg, David G. Robinson, and Harlan Yu. "Accountable Algorithms." *University of Pennsylvania Law Review* 165, no. 3 (2017): 633–706.

Lave, Jean. *Cognition in Practice: Mind, Mathematics and Culture in Everyday Life*. Cambridge: Cambridge University Press, 1988.

LeCun, Yann, Yoshua Bengio, and Geoffrey Hinton. "Deep Learning." *Nature* 521, no. 7553 (2015): 436–444. https://doi.org/10.1038/nature14539.

Lee, John D., and Katrina A. See. "Trust in Automation: Designing for Appropriate Reliance." *Human Factors* 46, no. 1 (2004): 50–80. https://doi.org/10.1518/hfes.46.1.50_30392.

Lu, Chris, Cong Lu, Robert Tjarko Lange, Jakob Foerster, Jeff Clune, and David Ha. "The AI Scientist: Towards Fully

Automated Open-Ended Scientific Discovery." arXiv, 2024. https://doi.org/10.48550/arXiv.2408.06292.

Luo, Renqian, Liai Sun, Yingce Xia, Tao Qin, Sheng Zhang, Hoifung Poon, and Tie-Yan Liu. "BioGPT: Generative Pre-trained Transformer for Biomedical Text Generation and Mining." *Briefings in Bioinformatics* 23, no. 6 (2022): bbac409. https://doi.org/10.1093/bib/bbac409.

March, James G. "Exploration and Exploitation in Organizational Learning." *Organization Science* 2, no. 1 (1991): 71–87. https://doi.org/10.1287/orsc.2.1.71.

March, James G., and Herbert A. Simon. *Organizations*. New York: Wiley, 1958.

McDuff, Daniel, Mike Schaekermann, Tao Tu, Anil Palepu, Amy Wang, Jake Garrison, and Karan Singhal, et al. "Towards Accurate Differential Diagnosis with Large Language Models." *Nature* 642, no. 8067 (2025): 451–457. https://doi.org/10.1038/s41586-025-08869-4.

McLuhan, Marshall. *Understanding Media: The Extensions of Man*. New York: McGraw-Hill, 1964.

Meyer, Marshall W., and Vipin Gupta. "The Performance Paradox." *Research in Organizational Behavior* 16 (1994): 309–369.

Mokyr, Joel. *The Lever of Riches: Technological Creativity and Economic Progress*. New York: Oxford University Press, 1990.

Munafò, Marcus R., Brian A. Nosek, Dorothy V. M. Bishop, Katherine S. Button, Christopher D. Chambers, Nathalie Percie du Sert, Uri Simonsohn, et al. "A Manifesto for Reproducible Science." *Nature Human Behaviour* 1 (2017): 0021. https://doi.org/10.1038/s41562-016-0021.

National Institute of Standards and Technology. *Artificial Intelligence Risk Management Framework (AI RMF 1.0)*. NIST AI 100-1. Gaithersburg, MD: National Institute of Standards and Technology, 2023. https://doi.org/10.6028/NIST.AI.100-1.

Newell, Allen, and Herbert A. Simon. *Human Problem Solving*. Englewood Cliffs, NJ: Prentice-Hall, 1972.

Norman, Donald A. *Things That Make Us Smart: Defending Human Attributes in the Age of the Machine*. Reading, MA: Addison-Wesley, 1993.

Norman, Donald A. *The Design of Everyday Things*. New York: Doubleday, 1988.

North, Douglass C. *Institutions, Institutional Change and Economic Performance*. Cambridge: Cambridge University Press, 1990.

Nosek, Brian A., George Alter, George C. Banks, D. Stephen Borsboom, Sara Bowman, Steven J. Breckler, Stuart Buck, et al. "Promoting an Open Research Culture." *Science* 348, no. 6242 (2015): 1422–1425. https://doi.org/10.1126/science.aab2374.

OECD. *Artificial Intelligence in Science: Challenges, Opportunities and the Future of Research*. Paris: OECD Publishing, 2023. https://doi.org/10.1787/a8d820bd-en.

OECD. *Governing with Artificial Intelligence: The State of Play and Way Forward in Core Government Functions*. Paris: OECD Publishing, 2025. https://doi.org/10.1787/795de142-en.

OECD. *Recommendation of the Council on Artificial Intelligence*. OECD/LEGAL/0449. Paris: OECD, 2019. Revised 2024.

https://legalinstruments.oecd.org/en/instruments/oecd-legal-0449.

Ong, Walter J. *Orality and Literacy: The Technologizing of the Word*. London: Routledge, 1982.

OpenAI. *GPT-4 Technical Report*. arXiv, 2023. https://doi.org/10.48550/arXiv.2303.08774.

Orlikowski, Wanda J. "Using Technology and Constituting Structures: A Practice Lens for Studying Technology in Organizations." *Organization Science* 11, no. 4 (2000): 404–428. https://doi.org/10.1287/orsc.11.4.404.14600.

Ostrom, Elinor. *Governing the Commons: The Evolution of Institutions for Collective Action*. Cambridge: Cambridge University Press, 1990.

Parasuraman, Raja, and Victor Riley. "Humans and Automation: Use, Misuse, Disuse, Abuse." *Human Factors* 39, no. 2 (1997): 230–253. https://doi.org/10.1518/001872097778543886.

Powell, Walter W. "Neither Market nor Hierarchy: Network Forms of Organization." *Research in Organizational Behavior* 12 (1990): 295–336.

Risko, Evan F., and Sam J. Gilbert. "Cognitive Offloading." *Trends in Cognitive Sciences* 20, no. 9 (2016): 676–688. https://doi.org/10.1016/j.tics.2016.07.002.

Rumelhart, David E., Geoffrey E. Hinton, and Ronald J. Williams. "Learning Representations by Back-Propagating Errors." *Nature* 323 (1986): 533–536. https://doi.org/10.1038/323533a0.

Rumelhart, David E., James L. McClelland, and the PDP Research Group. *Parallel Distributed Processing:*

Explorations in the Microstructure of Cognition. Volume 1: Foundations. Cambridge, MA: MIT Press, 1986.

Russell, Stuart. *Human Compatible: Artificial Intelligence and the Problem of Control.* New York: Viking, 2019.

Selbst, Andrew D., Danah Boyd, Sorelle A. Friedler, Suresh Venkatasubramanian, and Janet Vertesi. "Fairness and Abstraction in Sociotechnical Systems." In *Proceedings of the Conference on Fairness, Accountability, and Transparency*, 59–68. New York: Association for Computing Machinery, 2019. https://doi.org/10.1145/3287560.3287598.

Senge, Peter M. *The Fifth Discipline: The Art and Practice of the Learning Organization.* New York: Doubleday, 1990.

Simon, Herbert A. "The Architecture of Complexity." *Proceedings of the American Philosophical Society* 106, no. 6 (1962): 467–482.

Simon, Herbert A. *The Sciences of the Artificial.* Cambridge, MA: MIT Press, 1969.

Szymanski, Nathan J., Bernardus Rendy, Yuxing Fei, Rishi E. Kumar, Tanjin He, David Milsted, and Matthew J. McDermott, et al. "An Autonomous Laboratory for the Accelerated Synthesis of Inorganic Materials." *Nature* 624 (2023): 86–91. https://doi.org/10.1038/s41586-023-06734-w.

Terman, Lewis M. *The Measurement of Intelligence: An Explanation of and a Complete Guide for the Use of the Stanford Revision and Extension of the Binet-Simon Intelligence Scale.* Boston: Houghton Mifflin, 1916.

Tversky, Amos, and Daniel Kahneman. "Judgment under Uncertainty: Heuristics and Biases." *Science* 185, no. 4157

(1974): 1124–1131.
https://doi.org/10.1126/science.185.4157.1124.

Vygotsky, Lev S. *Mind in Society: The Development of Higher Psychological Processes*. Edited by Michael Cole, Vera John-Steiner, Sylvia Scribner, and Ellen Souberman. Cambridge, MA: Harvard University Press, 1978.

Weick, Karl E. *Sensemaking in Organizations*. Thousand Oaks, CA: Sage, 1995.

Wiggins, Grant. *Educative Assessment: Designing Assessments to Inform and Improve Student Performance*. San Francisco: Jossey-Bass, 1998.

Wilkinson, Mark D., Michel Dumontier, IJsbrand Jan Aalbersberg, Gabrielle Appleton, Myles Axton, Arie Baak, Niklas Blomberg, et al. "The FAIR Guiding Principles for Scientific Data Management and Stewardship." *Scientific Data* 3 (2016): 160018.
https://doi.org/10.1038/sdata.2016.18.

Winner, Langdon. "Do Artifacts Have Politics?" *Daedalus* 109, no. 1 (1980): 121–136.

Zhang, Jiajie. "The Nature of External Representations in Problem Solving." *Cognitive Science* 21, no. 2 (1997): 179–217. https://doi.org/10.1207/s15516709cog2102_3.

Zhang, Jiajie, and Donald A. Norman. "Representations in Distributed Cognitive Tasks." *Cognitive Science* 18, no. 1 (1994): 87–122.
https://doi.org/10.1207/s15516709cog1801_3

Zhang, Jiajie, and Vimla L. Patel. "Distributed Cognition, Representation, and Affordance." *Pragmatics & Cognition* 14, no. 2 (2006): 333–341.

Acknowledgments

This book was not written in a single place, or at a single time.

It emerged across years of thinking, teaching, leading, and building at the intersection of cognition, technology, and institutional change. It draws on a long arc of work—research papers, lectures, strategy, essays, and conversations—that gradually converged into the framework presented here.

I am grateful to the colleagues, students, collaborators, and friends whose questions sharpened these ideas and whose work continues to expand what it means to think together. Their contributions are not external to this book—they are part of the system from which it emerged. I am equally grateful to the broader communities of scholars and practitioners who have made cognition, design, and artificial intelligence a living field of inquiry.

This book was also written in interaction with artificial intelligence systems. These systems contributed to drafting, restructuring, and refining the work—generating alternatives and accelerating exploration. They were not authors, but participants in the process of thinking. The ideas, interpretations, and conclusions are entirely my own.

In this sense, the book is not only about the cognitive revolution—it is written within it.

Finally, I thank my family for their patience, support, and unwavering belief throughout the long arc of this work. They have carried the weight of this project in ways that do not appear on the page—through time given, attention deferred, and the quiet endurance required for sustained intellectual work. In a book about how thinking is distributed, it is only fitting to acknowledge that this work, too, was sustained by others.

This book was not written alone.

About the Author

Jiajie Zhang, PhD, is a cognitive scientist, dean, and professor whose work explores distributed cognition, external representations, human–AI interaction, and the future of institutional design.

Across scholarship, leadership, and strategy, his work focuses on how intelligence is moving from individual minds into systems of people, tools, and machines. The Cognitive Revolution brings these themes together into a unified framework for rethinking expertise, education, research, governance, leadership, and the design of AI-native institutions.